Radar Sensors: From Cruise Control to Safety

Shiva

Contents

Chapter 1

Introduction

1.1 Motivation

Radar systems have become widely employed in automotive applications over the last few years. The digital automotive radar is particularly flexible, where the frequency-modulated continuous wave-based radar exhibit superior properties considering the requirements for the environment perception task. Since the early 2000s, the radar sensor is used in automotive applications, e.g., adaptive cruise control. Back then, both simultaneous high-resolution target range and velocity measurement in multi-target situations was a critical requirement for such applications. The well-known waveforms had a considerably long acquisition time from $50-100\,\mathrm{ms}$ [RM01]. Continuous progress in the development of the frequency modulated continuous wave (FMCW) chirp sequence technique reduced the acquisition time for a single chirp to the sub-milliseconds region. The ability to provide three-dimensional target information consisting of range, radial velocity and azimuthal direction in poor visibility and weather conditions make radar sensors indispensable in addressing the challenges of autonomous driving. The robustness to adverse weather conditions [HKDB16, HDKR17] and the low costs led to large-scale integration of radar sensors in intelligent transportation systems [DKH$^+$15]. FMCW radar systems have successfully been applied to realize automotive applications such as blind-spot detection, lane change assistance, smart cruise control, parking and anti-collision warning systems. At this time, low-cost FMCW radars are preferred in such automotive applications [JJL14].

However, the environment perception data may also be used to further enhance the robustness or extend the functionality of passive or integral safety functions [CCL11, RCY17]. In case a collision is unavoidable, passive or integral safety systems trigger reversible and irreversible restraint systems to mitigate occupant injury or fatalities based on acceleration and pressure sensor information. Today, these sensors rely on the contact between the collision partners in order to measure, e.g., a deceleration and are thus able to execute safeguards after this initial contact moment.

The temporal activation moment of restraint systems may be shifted towards or even before the contact moment by incorporating observations obtained by forward-looking sensors. Observing the local environment with forward-looking sensors and utilizing the information for passive or integral safety applications migrates the high requirements for activating restraint systems. Extensive requirements, e.g., with minimal misfire rates, are necessary since the false activation of restraint systems may restrict the driver from correctly executing the driving task and may place oneself and others in dangerous or life-threatening situations. Hence, environment perception for future safety systems requires high accuracy to minimize false activation of safety systems. The introduction of high-resolution radar sensor perceiving short-range targets and the utilization for safety applications generate demand for innovations to be developed and opened up the research area to further enhance automotive safety systems.

This work focuses on radar-based environment perception in pre-crash scenarios. The following chapters describe novel methods to detect and track potential collision targets, which are in short-range distances and on potential collision trajectories or execute an evading maneuver, as well as a method to reproducibly test these systems. Moreover, the problem of automotive multipath propagation in uncertain environments is considered.

1.2 Book overview

In the introducing chapter, the motivation and initial problems for the proposed methods are presented. This section introduces each chapter and the original contributions are summarized.

The second chapter covers the fundamentals of electromagnetic wave propagation and gives a comprehensive derivation of the FMCW chirp sequence technique. The experimental FMCW radar system and raw signal processing are introduced. For the majority of this work, a configurable FMCW radar system is used to conduct high-resolution radar measurements.

In the third chapter, a description of the Fresnel and Fraunhofer region and definitions for radar target detections in short-range distances are given. Moreover, the chapter provides a theoretical derivation of the micro-Doppler effect and connects the effect with the radar sensor by introducing exemplary high-resolution radar data of potential collision targets. In particular, the micro-Doppler effect generated by rotating components, e.g., spinning wheels, is measured and evaluated to acquire additional, valuable information about the target object. A method is proposed to determine the position and corresponding bulk velocity of the target spinning wheels, which serve as fixed and characteristic points on the vehicle surface. This novel information serves as input for a subsequent target state estimation procedure in the following chapter.

In comparison to the signals scattered from the object of interest directly back to the sensor, target vehicle originated signals in close distance may be subject to multipath propagation. These multipath signals are referred to as false-positive detections. The transmitted electromagnetic wave gets potentially reflected by other objects or structure surfaces like, e.g., guardrails. The chapter addresses this problem by introducing a method to identify potential false-positive detections based on multipath propagation.

Chapter four opens with a summary of a vehicle collision study, which revealed skid driving situations as frequent and crucial pre-crash scenarios. These results motivated the development of a reproducible and non-destructive test method in skid scenarios to evaluate the performance of mounted environment sensors. Extensive real vehicle tests using the novel test method motivated the design of a target tracking system in skid driving situations. The ego-vehicle horizontal movement is estimated using a novel motion model. This model considers an additional lateral force to estimate a characteristic dynamic parameter, the side slip angle. The parameter serves as input to the designed filter stage.

Moreover, the chapter provides a brief but comprehensive overview of Bayesian state estimation and introduces three fundamental filter designs: the Kalman filter (KF), the extended Kalman filter (EKF) and the particle filter (PF). A subsequent particle-filter stage incorporates the previously detected target spinning wheel hypotheses is designed. The presented filter stage covers the signal processing from raw radar sensor data to extended target state estimation in pre-crash scenarios. The particle filter is evaluated on real experimental data where a target vehicle with a mounted reference sensor is used to reconstruct safety-critical, dynamic evading maneuvers.

Chapter five summarizes the scientific contributions, emphasizes their practical value for the proposed methods and provides an outlook for future work.

1.3   Original contributions

High-resolution, short-range, radar-based detection and tracking of target vehicles, that may be on a collision trajectory with the ego-vehicle, demand contrary requirements to state estimation than conventional environment perception. Generally speaking, the main differences are the number of objects is drastically lower, ego and target vehicle may be in high-dynamic or skid driving situations due to potential evading maneuvers and targets may cause large numbers of detections per object. A summary of the problems addressed in this **book** and original contributions coping with them is given in the following:

- *Wandering dominant scatter point on the target surface and subsequent state estimation:* In this context, the dominant scatter point (P_{DS}) is considered as the target detection with the maximal intensity. The position of the dominant scatter point strongly depends on the target relative yaw angle and the angle under which the sensor illuminates the target. As the dominant scatter point moves along the vehicle surface, an additional uncertainty in detection and tracking systems needs to be considered [KYY+15]. Fig. 1.1 shows three radar snapshots during an approaching sequence illustrating the wandering dominant scatter point on the vehicle surface. The colored circle indicates the fixed virtual reference point (P_{ref}) at the center of the target front bumper. The target vehicle approaches the sensor on a zigzag trajectory and subsequently pass the sensor. The Euclidean offset values range up to 1.2 m for snapshot t=52 and t=76 and drastically increase when the target is only partially within the sensor field of view at t=100. These additional uncertainty needs to be considered and require an appropriate state estimation method to achieve high accuracy requirements, which is needed in safety applications since crash severity estimation depend on the vehicle overlap [SB19].

The book at hand addresses this problem and proposes a method to determine rotating target wheel positions and bulk velocities, which serve as characteristic points on the vehicle surface and are input to a subsequent state estimation procedure.

- *Multipath propagation in pre-crash scenarios:* Another challenge is the type of propagation that the radar signal might take on the transmitting or receiving path, where radar receives not only the direct reflection of an obstacle but also indirect temporally shifted reflection components from potential reflection

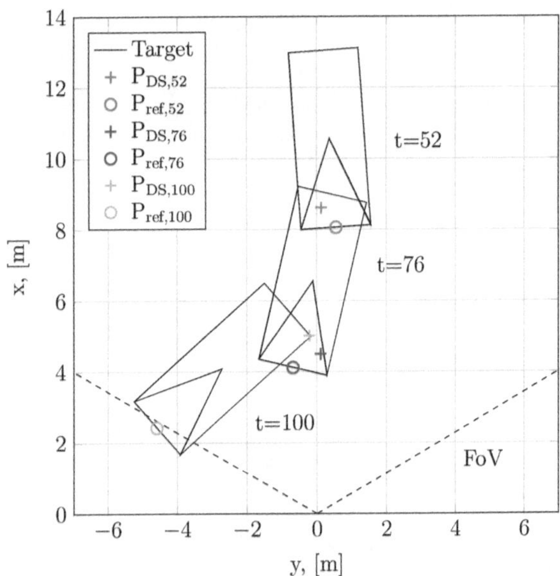

Figure 1.1: Three radar snapshots showing an illuminated target during an approaching sequence quantifying the wandering dominant scatter point (P_{DS}) on the vehicle surface. The colored circle indicates a fixed reference point (P_{ref}) on the target front bumper center. The target vehicle approaches the sensor on a zigzag trajectory, causing yaw angle variations and eventually pass the sensor. The Euclidean offset values range up to 1.2 m for snapshot t=52 and t=76 and drastically increase when the target is partially within the sensor field of view at t=100.

surfaces leading to false-positive detections [VHZ18, EHZ$^+$17]. The sensor deterioration creates high demands on signal-processing algorithms concerning the suppression of mispredictions. Thus, the radar receives not only the direct reflections of an obstacle but also indirect temporally delayed reflection components. This superposition can lead to range-dependent interference patterns, which cause oscillating signal amplitudes of the received power [DKS$^+$11]. In addition to performance degradation of the direction of arrival (DoA) estimation due to fading effects, multipath propagation can lead to the appearance of mirrored ghost targets [VHZ18, EHZ$^+$17, KHP$^+$18].

To solve this problem, a geometric model is proposed to identify potential multipath reflections and mirrored ghost targets.

- *Target tracking in skid driving pre-crash scenarios:* A conducted german in-depth accident study (GIDAS) proved relevance for unstable and skid driving situations in the pre-crash phase of a collision. According to the study, in 2015 occurred 754 collisions with injured occupants in Germany, where skid driving situations were present [Fei16]. The majority of them were caused by speeding in combination with inappropriate driver reactions.

  For this purpose, a non-destructive test method to evaluate forward-looking sensors in skid driving situations is developed and presented. The test method reproducibly transfers the test vehicle in skid driving situations. Subsequently, a survey using a test vehicle with a mounted state-of-the-art automotive radar sensor is carried out. The sensor detects and tracks a standardized target object during skid driving situations. Results revealed poor accuracy and drastically decreased detection reliability [KHD$^+$17, KBH$^+$17].

  As a solution to this problem, an EKF-based tracking procedure is designed to improve static object tracking in various skid driving situations. The tracker utilizes a novel motion model, which considers an additional lateral force to estimate dynamic parameters of the ego-vehicle.

A summary of the articles, which have been published during the period of doctoral candidacy, is given in the following.

Internationally refereed journal articles

- **A. Kamann**, D. Steinhauser, F. Gruson, T. Brandmeier, and U. T. Schwarz. Extended object tracking using spatially resolved micro-doppler signatures. In *2020 IEEE Transactions on Intelligent Vehicles*, in press, IEEE, 2020.

Internationally refereed conference proceedings

- **A. Kamann**, P. Held, F. Perras, P. Zaumseil, T. Brandmeier, and U. T. Schwarz. Automotive radar multipath propagation in uncertain environments. In *2018 21st International Conference on Intelligent Transportation Systems (ITSC)*, pages 859-864. IEEE, 2018.

- **A. Kamann**, J. B. Bielmeier, S. Hasirlioglu, U. T. Schwarz, and T. Brandmeier. Object tracking based on an extended Kalman filter in high dynamic driving situations. In *2017 IEEE 20th International Conference on Intelligent Transportation Systems (ITSC)*, pages 1-6. IEEE, 2017.

- **A. Kamann**, S. Hasirlioglu, I. Doric, T. Speth, T. Brandmeier, and U. T. Schwarz. Test methodology for automotive surround sensors in dynamic driving situations. In *2017 IEEE 85th Vehicular Technology Conference (VTC Spring)*, pages 1-6. IEEE, 2017.

- P. Held, D. Steinhauser, A. Kamann, A. Koch, T. Brandmeier and U. T. Schwarz. Normalization of micro-Doppler spectra for cyclists using high-resolution projection technique. In *IEEE International Conference on Vehicular Electronics and Safety (ICVES) 2019*. **Award: Runner-up for Best Paper Award.**

- P. Zaumseil, D. Steinhauser, P. Held, A. Kamann, and T. Brandmeier. Radar-based environment perception using back projection algorithm. In *European Radar Conference 2019*.

- D. Steinhauser, P. Held, A. Kamann, A. Koch, and T. Brandmeier. Micro-Doppler extraction of pedestrian limbs for high resolution automotive radar. In *2019 IEEE Intelligent Vehicles Symposium (IV)*, pages 764-769. IEEE, 2019.

- P. Held, D. Steinhauser, A. Kamann, A. Koch, T. Brandmeier, and U. T. Schwarz. Micro-Doppler extraction of bicycle pedaling movements using automotive radar. In *2019 IEEE Intelligent Vehicles Symposium (IV)*, pages 744-749. IEEE, 2019.

- P. Held, D. Steinhauser, A. Kamann, T. Holdgrün, I. Doric, A. Koch, and T. Brandmeier. Radar-based analysis of pedestrian micro-Doppler signatures using motion capture sensors. In *2018 IEEE Intelligent Vehicles Symposium (IV)*, pages 787-793. IEEE, 2018.

- S. Hasirlioglu, I. Doric, A. Kamann, and A. Riener. Reproducible fog simulation for testing automotive surround sensors. In *2017 IEEE 85th Vehicular Technology Conference (VTC Spring)*, pages 1-7. IEEE, 2017.

- S. Hasirlioglu, A. Kamann, I. Doric, and T. Brandmeier. Test methodology for rain influence on automotive surround sensors. In *2016 IEEE 19th International Conference on Intelligent Transportation Systems (ITSC)*, pages 2242-2247. IEEE, 2016.

- T. Speth, A. Kamann, T. Brandmeier, and U. Jumar. Precise relative ego-positioning by stand-alone RTK-GPS. In *2016 13th Workshop on Positioning, Navigation and Communications (WPNC)*, pages 1-6. IEEE, 2016.

- K. Schneider, G. J. Sequeira, R. Lugner, A. Kamann T. Brandmeier, R. Burgmeier. Verification of pre-crash information for a predictive activation of passive safety actuators. In *11. VDI Tagung Fahrzeugsicherheitstechnik*, VDI-Berichte 2312, ISBN 978-3-18-092312-3.

Published patent applications

- Steuervorrichtung und Verfahren zum Ermitteln der Eigenposition eines Kraftfahrzeugs. Erfinder: T. Speth, T. Brandmeier, A. Kamann; Anmelder: Audi AG; Anmeldedatum; 08.03.2016; Anmeldeaktenzeichen: DE102016002704.

- Aufprallsensor mit zumindest zwei voneinander beabstandeten Elektroden sowie Verfahren zur Aufprallerkennung sowie Auslösung von Schutzeinrichtungen mit einem solchen Aufprallsensor. Erfinder: T. Tyroller, R. Burgmeier, K. Schneider, G. J. Sequeira, T. Brandmeier, A. Kamann, R. Lugner; Anmelder: Continental AG; Anmeldedatum; 15.01.2018; Anmeldeaktenzeichen: DE1020-18200604.

- Verfahren zum Erzeugen eines Umfeldmodells für ein Fahrzeug. Erfinder: A. Kamann, T. Brandmeier, M. Feser, B. Peters, A. Forster; Anmelder: Continental AG; Anmeldedatum; 13.10.2017; Anmeldeaktenzeichen: DE102017218340.

- Verfahren zum Ermitteln einer Anzahl von Daten eines entfernten Fahrzeugs. Erfinder: R. Naumann, A. Kamann, F. Perras, T. Brandmeier, A. Forster; Anmelder: Continental AG; Anmeldedatum; 26.04.2018; Anmeldeaktenzeichen: DE102018206500.

- Konturerkennung eines Fahrzeugs anhand von Messdaten einer Umfeldsensorik. Erfinder: M. Feser, K. Schneider, T. Brandmeier, A. Kamann, R. Lugner, D. Weichselberger, R. Naumann; Anmelder: Continental AG; Anmeldedatum; 02.05.2018; Anmeldeaktenzeichen: DE102018206500 and PCT/-DE2019/200031.

Unpublished patent applications

- Verfahren zur Bestimmung von unfallrelevanten Parametern mittels eines Radarsystems eines Fahrzeugs. Erfinder: A. Kamann, A. Koch, T. Brandmeier, F. Gruson, P. Held, A. Forster; Anmelder: Continental AG; Anmeldedatum; 24.08.2018; Anmeldeaktenzeichen: DE102018214338 and PCT/DE2019/2000-27.

- Verfahren zum Erfassen von Teilbereichen eines Objekts. Erfinder: D. Steinhauser, P. Held, A. Kamann, T. Brandmeier, A. Koch, A. Abdulrahman; Anmelder: Continental AG; Anmeldedatum; 08.01.2019; Anmeldeaktenzeichen: DE102019200141.

- Verfahren zum Erfassen einer Bewegung eines Objekts. Erfinder: P. Held, D. Steinhauser, A. Kamann, T. Brandmeier, A. Koch, A. Abdulrahman; Anmelder: Continental AG; Anmeldedatum; 24.01.2019; Anmeldeaktenzeichen: DE102019200849.

- Verfahren zum Verfolgen eines entfernten Zielfahrzeugs in einem Umgebungsbereich eines Kraftfahrzeugs mittels einer Kollisionserkennungsvorrichtung. Erfinder: A. Kamann, R. Naumann, D. Steinhauser, T. Brandmeier, A. Forster, H. Faisst, F. Gruson; Anmelder: Continental AG; Anmeldedatum; 31.10.2019; Anmeldeaktenzeichen: DE102019216805.

References

[CCL11] Kwanghyun Cho, Seibum B. Choi, and Hyeongcheol Lee. Design of an airbag deployment algorithm based on precrash information. *IEEE Transactions on vehicular technology*, 60(4):1438–1452, 2011.

[DKH⁺15] Jürgen Dickmann, Jens Klappstein, Markus Hahn, Marc Muntzinger, Nils Appenrodt, Carsten Brenk, and Alfons Sailer. Present research activities and future requirements on automotive radar from a car manufacturer's point of view. In *2015 IEEE MTT-S International Conference on Microwaves for Intelligent Mobility (ICMIM)*, pages 1–4. IEEE, 2015.

[DKS⁺11] Fabian Diewald, Jens Klappstein, Frederik Sarholz, Jürgen Dickmann, and Klaus Dietmayer. Radar-interference-based bridge identification for collision avoidance systems. In *2011 IEEE Intelligent Vehicles Symposium (IV)*, pages 113–118. IEEE, 2011.

[EHZ⁺17] Florian Engels, Philipp Heidenreich, Abdelhak M. Zoubir, Friedrich K. Jondral, and Markus Wintermantel. Advances in automotive radar: A framework on computationally efficient high-resolution frequency estimation. *IEEE Signal Processing Magazine*, 34(2):36–46, 2017.

[Fei16] Harald Feifel. Crash constellation and severity for vehicle-to-vehicle collisions with presence of side slip angles. In *Internal GIDAS query*. Continental AG, 2016.

[HDKR17] Sinan Hasirlioglu, Igor Doric, Alexander Kamann, and Andreas Riener. Reproducible fog simulation for testing automotive surround sensors. In *2017 IEEE 85th Vehicular Technology Conference (VTC Spring)*, pages 1–7. IEEE, 2017.

[HKDB16] Sinan Hasirlioglu, Alexander Kamann, Igor Doric, and Thomas Brandmeier. Test methodology for rain influence on automotive surround sensors. In *2016 IEEE 19th International Conference on Intelligent Transportation Systems (ITSC)*, pages 2242–2247. IEEE, 2016.

[JJL14] Yeonghwan Ju, Youngseok Jin, and Jonghun Lee. Design and implementation of a 24 GHz FMCW radar system for automotive applications. In *2014 International Radar Conference*, pages 1–4. IEEE, 2014.

[KBH⁺17] Alexander Kamann, Jonas B. Bielmeier, Sinan Hasirlioglu, Ulrich T. Schwarz, and Thomas Brandmeier. Object tracking based on an extended Kalman filter in high dynamic driving situations. In *2017 IEEE 20th International Conference on Intelligent Transportation Systems (ITSC)*, pages 1–6. IEEE, 2017.

[KHD+17] Alexander Kamann, Sinan Hasirlioglu, Igor Doric, Thomas Speth, Thomas Brandmeier, and Ulrich T. Schwarz. Test methodology for automotive surround sensors in dynamic driving situations. In *2017 IEEE 85th Vehicular Technology Conference (VTC Spring)*, pages 1–6. IEEE, 2017.

[KHP+18] Alexander Kamann, Patrick Held, Florian Perras, Patrick Zaumseil, Thomas Brandmeier, and Ulrich T. Schwarz. Automotive radar multipath propagation in uncertain environments. In *2018 21st International Conference on Intelligent Transportation Systems (ITSC)*, pages 859–864. IEEE, 2018.

[KYY+15] Beomjun Kim, Kyongsu Yi, Hyun-Jae Yoo, Hyok-Jin Chong, and Bongchul Ko. An IMM/EKF approach for enhanced multitarget state estimation for application to integrated risk management system. *IEEE Transactions on Vehicular Technology*, 64(3):876–889, 2015.

[RCY17] Manoharprasad K. Rao, Mark Cuddihy, and Wilford Yopp. Method for operating a pre-crash sensing system to deploy airbags using confidence factors prior to collision, May 30 2017. US Patent 9,663,052.

[RM01] Hermann Rohling and M. Meinecke. Waveform design principles for automotive radar systems. In *2001 CIE International Conference on Radar Proceedings (Cat No. 01TH8559)*, pages 1–4. IEEE, 2001.

[SB19] Gerald J. Sequeira and Thomas Brandmeier. Evaluation and characterization of crash-pulses for head-on collisions with varying overlap crash scenarios. *Transportation Research Procedia, in press*, 2019.

[VHZ18] Tristan Visentin, Jürgen Hasch, and Thomas Zwick. Analysis of multipath and DOA detection using a fully polarimetric automotive radar. *International Journal of Microwave and Wireless Technologies*, 10(5-6):570–577, 2018.

Chapter 2

Automotive radar

This chapter introduces the general electromagnetic wave propagation principles developed by Friis, presents a experimental radar sensor system and gives a comprehensive derivation of fundamentals of frequency modulated continuous wave (FMCW) fast chirp sequence technique for automotive environment perception. For the majority of this work, a configurable FMCW radar system is used to conduct real radar measurements. To further enhance the target detection, a thresholding and clustering procedure is presented, discussed and exemplary applied to real radar data.

2.1 Electromagnetic wave propagation

The isotropic electromagnetic wave propagation in free space between a transmitting and receiving antenna can be modeled by Friis' equations [Fri46]. The effective antenna area, whether transmitting or receiving, to receive a linearly polarized, plane electromagnetic wave is defined by

$$A_{\textit{eff}} = \frac{P_r}{P_0},$$

(2.1)

where P_r is the power available at the output terminals of the receiving antenna and P_0 is the power flow per unit area of the incident field at the antenna. Considering a setup with an isotropic transmitting antenna and a receiving antenna with effective area A_r, the power flow per unit area at a distance d from the transmitter is

$$P_r = A_{\textit{eff}} P_0 = A_{\textit{eff}} \frac{P_t}{4\pi d^2},$$

(2.2)

where P_t is power fed into the transmitting antenna at its input terminals. Replacing the isotropic transmitting antenna with a transmitting antenna with effective area A_t will increase the received power by the ratio A_t/A_{isotr} with $A_{isotr} = \lambda^2/4\pi$ and one obtains

$$\frac{P_r}{P_t} = \frac{A_r A_t}{d^2 \lambda^2},$$

(2.3)

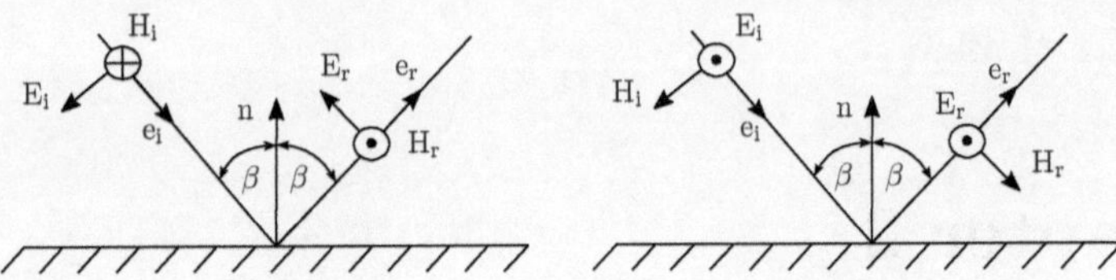

Figure 2.1: Reflection geometry and orientation of electric field strength of an incident and reflected plane wave for parallel (left) and orthogonal (right) polarization ©2018 IEEE [KHP+18].

where λ is the wavelength. The relation between an antenna gain and the effective antenna area can be described by [Rem16]

$$G_{r,t} = \frac{4\pi}{\lambda^2} A_{r,t},$$ (2.4)

where G_r is the antenna receiver gain and G_t the antenna transmitter gain. Thus, the received power between a transmitting and receiving antenna in a distance d can be expressed by

$$\frac{P_r}{P_t} = G_r G_t \left(\frac{\lambda}{4\pi d} \right)^2.$$ (2.5)

In automotive environment perception applications, the electromagnetic wave is likely to get reflected on obstacles or surfaces in the sensor field of view (FoV) and travels back to the sensor.

In real-world environments, where, e.g., buildings, other vehicles or barriers are present, the transmitted energy gets either absorbed, transmitted through the material, reflected (specular reflection) or scattered. The reflectivity and absorption properties of the object depend on the polarization of the electromagnetic wave. The polarization is defined as either parallel or orthogonal to the reflection plane generated by the direction of incidence. For parallel polarization, the incident radiation of the E-field is oriented parallel to the plane of incidence, and for perpendicular polarization, the E-field is perpendicular to the plane of incidence. For vertical obstacle surfaces, e.g., other cars or walls and horizontal propagation of the radar beam, horizontal polarization is parallel to the plane of incidence while vertical polarization is perpendicular.

Fig. 2.1 shows the corresponding geometric relations. The reflection coefficient depends on conductivity, angle of incidence, permittivity, roughness of the surface and the polarization of the incident wave. A general reflection coefficient R can be defined by

$$R = R_s \cdot \rho,$$ (2.6)

where R_s is the smooth surface reflection coefficient and ρ is the scalar surface roughness attenuation factor. The parallel reflection coefficient $R_{s\parallel}$ and perpendicular reflection coefficient $R_{s\perp}$ are given by the Fresnel equations

$$R_{s\parallel} = \frac{\sin\beta - \sqrt{\epsilon - \cos^2\beta}}{\sin\beta + \sqrt{\epsilon - \cos^2\beta}}, \tag{2.7}$$

$$R_{s\perp} = \frac{\epsilon\sin\beta - \sqrt{\epsilon - \cos^2\beta}}{\epsilon\sin\beta + \sqrt{\epsilon - \cos^2\beta}}, \tag{2.8}$$

where β is the angle of incidence and ϵ is the complex permittivity given by

$$\epsilon = \epsilon_0\epsilon_r - j\frac{\sigma}{\omega_{\mathrm{Refl}}}, \tag{2.9}$$

where ϵ_0 is the permittivity of vacuum, ϵ_r is the relative dielectric constant, σ the conductivity of the reflecting surface and ω_{Refl} the angular frequency. The roughness attenuation factor for ρ is given as [CL82]

$$\rho^2 = e^{-2\delta}, \quad \text{with } \delta = \frac{4\pi\Delta h}{\lambda}\sin\beta, \tag{2.10}$$

where Δh is the standard deviation of the normal distribution of the surface roughness.

In real-world environments, the particular roughness features are usually vaguely known and depend on obstacles geometries. Thus, objects classification may be realized by evaluating the radar cross section (RCS) parameter, a property of a radar target to represent the echo signal returned to the radar sensor [HAS08]. The RCS parameter is defined for large d $(d \to \infty)$ as

$$\sigma = \lim_{d\to\infty}\left\{4\pi d^2\frac{|E_s|^2}{|E_i|^2}\right\}, \tag{2.11}$$

where $|E_s|$ and $|E_i|$ are the scattered and incident electric fields, respectively. The RCS can be considered as the objects effective scattering cross section, which also depends on the relative azimuthal sensor-to-object configuration and the wavelength λ. This value can also serve as a simple obstacle identifier. Assuming small RCS values for small targets, e.g., pedestrians, and larger RCS values for larger objects, e.g., cars or trucks.

Hence, the final received and transmitted power ratio can be expressed, considering the RCS as effective area and twice the sensor to object electromagnetic wave travel distance, as

$$\frac{P_r}{P_t} = G_r G_t\frac{\lambda^2\sigma}{(4\pi)^3 d^4}, \tag{2.12}$$

for automotive radar applications.

2.2 FMCW radar system

To the best of the author's knowledge, the first recorded idea to utilize a radar system in the automotive safety context appeared in 1955 [Fon55] and describes a safety system to decelerate the vehicle based on radar environment information. It took four more decades to realize the first driver assistance function in series-produced vehicles where the radar system was adaptively controlling cruise speed depending on vehicles in the local environment [WHW09]. Since then, the number of vehicles and driver assistance systems, including the number of radar sensors, grew tremendously.

Nowadays, low-cost FMCW radars are preferred in automotive applications [JJL14]. Vehicles are equipped with multiple radar sensors employing data fusion. Thus, multiple radar sensors cover a wider sensor FoV than a single radar sensor can capture and give a comprehensive overview of the surrounding scene. According to [WHW09], one can group radar functions in *comfort* and *safety* applications.

Comfort applications aim to minimize the workload of the driving task by partially or fully automating the vehicle control, e.g., adaptive cruise control (ACC), where the system sets the cruising speed to the desired level or adapts it to the actual traffic situation. Next generations of comfort applications intend to maximize the vehicle driving control towards fully-automated driving where the system is capable of handling the vehicle in every driving situation.

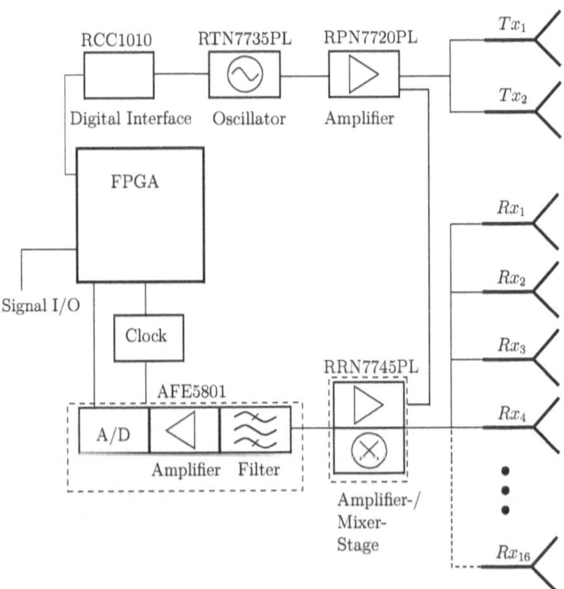

Figure 2.2: Hardware configuration for a FMCW radar system with two transmit antennas (Tx) and multiple receive antennas (Rx).

Safety applications categorize into active safety and passive safety, where the first group attempts to avoid collisions by longitudinal and lateral intervention in vehicle control. The latter group of system mitigates crash severity, e.g., occupant injuries, of unavoidable collisions. These systems trigger cascaded warnings and eventually activate reversible and irreversible restraints systems, e.g., belt tensioners or airbags in various applications, e.g., emergency brake assist.

2.2.1 Radar system platform

The presented FMCW-radar block diagram is based on the INRAS technical reports and manuals of the experimental radar sensor [Had19]. Fig. 2.2 shows the main components of the experimental FMCW radar. It consists of two transmit antennas and 16 receive antennas implemented as a patch antenna array. The 16 receive antennas form a uniform linear array (ULA) with element spacing d_{ant}. The field programmable gate array (FPGA) controls each component and serves as an interface to the measurement computer via its signal input/output port forwarding recorded radar data. An oscillator (RTN7735), accessed through a digital interface (RCC1010), including a phased-locked loop (PLL) frequency control in conjunction with a dual power amplifier stage (RPN7720PL), generates the arbitrary chirp sequences as FMCW transmit signal. The activation sequence can be programmed with the trigger and timing unit implemented in the FPGA. The receive path is realized with four receiver components (RRN7745) connecting the 16 receiving antennas. The incident electromagnetic wave is captured by the receiving antennas, down-converted to the base-band and amplified (RRN7745PL). Subsequently, the incoming signal is filtered, amplified again and sampled using an analog-digital converter (ADC) with an adjustable sampling frequency (AFE5801). Table 2.1 shows system parameters for an exemplary short range radar (SRR) configuration. Experimental results using these or similar settings are presented in the following sections.

Table 2.1: Exemplary radar configuration

Parameter	Value
Carrier frequency f_c	77 GHz
Bandwidth B	2 GHz
Sampling frequency	10 MHz
Chirp duration	51.2 µs
Samples per chirp	512
Number of chirps per interval	512
Chirp repetition interval	60 µs
Range resolution	0.075 m
Doppler resolution	$0.06\,\frac{m}{s}$

2.2.2 Chirp sequence data model

The description of the FMCW radar chirp sequence measurement model is based on the work of Costa *et al.* [CC84] and Barrick *et al.* [Bar73]. In their contributions, the authors presented the fast chirp sequence FMCW method and how, both time-delay (range) and Doppler (radial velocity) information, can be extracted unambiguously.

The waveform emitted by the transmitting antenna, generated from a voltage controlled oscillator, is a linear frequency modulated pulse referred to as *sweep* or *chirp*. The signal can be described for $-T_r/2 < t < T_r/2$ as

$$E_T(t) = E_T^\circ \cos\left[\phi_T(t)\right], \quad \text{with } \phi_T(t) = 2\pi f_c t + \pi B f_r t^2, \tag{2.13}$$

where E_T° is the amplitude, f_c is the carrier frequency, f_r is the pulse repetition frequency, $T_r = 1/f_r$ is the chirp duration, B is the bandwidth, and ϕ_T is the transmitting signal phase. The signal is phase-coherent and periodic from one repetition interval to the subsequent. The instantaneous frequency $f_T(t)$ is the derivative of the phase and therefore

$$f_T(t) = \frac{1}{2\pi}\frac{d\phi_T(t)}{dt} = f_c + B f_r t. \tag{2.14}$$

Fig. 2.3 shows the transmitted, successive frequency chirps and the delayed and Doppler-shifted received signal due to a distant, moving target. The target range $R(t)$ is a function of time as $R(t) = R_0 + vt$ where v is the target velocity along line of sight (LoS). The received signal, scattered back from target to the sensor, is a replica of the transmitted signal delayed in time by t_d with $t_d = 2R/c$, where c is the speed of light. If the target is moving, its frequency is shifted accordingly. The received signal, shown in Fig. 2.3 as the dashed curve, can be expressed as

$$E_R(t) = E_T^\circ(t - t_d) = A_R \cos\left[2\pi f_c(t - t_d) + \pi B f_r(t - t_d)^2\right], \tag{2.15}$$

where A_R is the amplitude of the echo signal. The received signal is mixed with a replica of the transmitted signal, which is mathematically represented by a multiplication of both signals. Using the addition theorem

$$\cos(\phi_T(t)) \cdot \cos(\phi_T(t - t_d)) =$$
$$\frac{1}{2}\left[\cos(\phi_T(t) - \phi_T(t - t_d)) + \cos(\phi_T(t) + \phi_T(t - t_d))\right], \tag{2.16}$$

where the first term produces the signal of interest at low frequencies while the second term contains high frequencies and is therefore filtered by the bandpass. Thus, the term $\cos(\phi_T(t) - \phi_T(t - t_d))$ is left for analysis. The signal with all phase terms is

$$x_{IF}(t) = X_{IF} \cos\left[2\pi f_c t - 2\pi f_c(t - t_d) + \pi B f_r t^2 - \pi B f_r(t - t_d)^2\right], \tag{2.17}$$

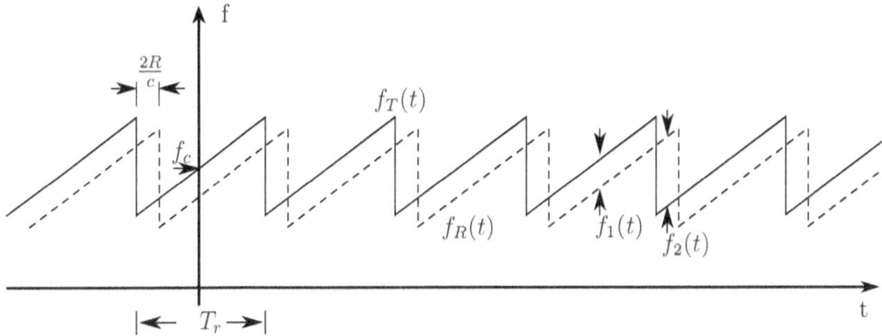

Figure 2.3: Transmitted and received signal, which is delayed in time and Doppler-shifted, due to reflection on a distant, moving target.

where X_{IF} is the amplitude of the mixed signal. The mixture of the two frequencies produces a dechirped, intermediate signal $x_{IF}(t)$. The frequency $f_I(t)$, shown in Fig. 2.4, consists of the two frequencies indicated in Fig. 2.3:

$$f_1 = \frac{1}{2\pi}\frac{d}{dt}\left[\phi_T(t - t_d) - \phi_T(t)\right], \tag{2.18}$$

$$f_2 = \frac{1}{2\pi}\frac{d}{dt}\left[\phi_T(t - t_d) - \phi_T(t + T_r)\right]. \tag{2.19}$$

The intermediate signal can be represented as the sum of pulse sequences as shown in Fig. 2.4. The signal $x_1(t)$ is at frequency f_1 with pulse width $T = T_r - t_d$ and $x_2(t)$ is at frequency f_2 with pulse width $T = t_d$. The latter frequency will be filtered since $f_2 \gg f_1$ and is therefore not of interest.

After a re-centering the time origin to the middle of the first pulse, the derivation continues within a chirp interval $-T_r/2 < t < T_r/2$. Since the target is moving, the frequency and phase from pulse to pulse is slightly changing which is key to the subsequent velocity derivation. The inter chirp time is denoted as t_i. Further phase simplifications for the first pulse are $t = t_i$ and $t_d = 2R/c = 2R_0/c + 2vt/c = t_0 + 2vt_i/c$ where $t_0 = 2R_0/c$ is the initial delay of the signal.
Hence, the phase of the down-converted and filtered signal $x_{IF}(t)$ is $\phi_I(t_i) = \phi_T(t_i - t_d) - \phi_T(t_i)$ or

$$\begin{aligned}
\phi_I(t_i) =& \left(-2\pi f_c t_0 + \pi B f_r t_0^2\right) \\
&+ 2\pi\left[-2\left(\frac{v}{c}\right)f_c + B f_r t_0\left(\frac{2v}{c}\right) - B f_r t_0\right]t_i \\
&- 2\pi B f_r\left(\frac{2v}{c}\right)\left[1 - \left(\frac{v}{c}\right)\right]t_i^2,
\end{aligned} \tag{2.20}$$

with three contributions to the phase. The quadratic term t_i^2 can be neglected since it is small with respect to the linear and constant terms, if one applies radar settings

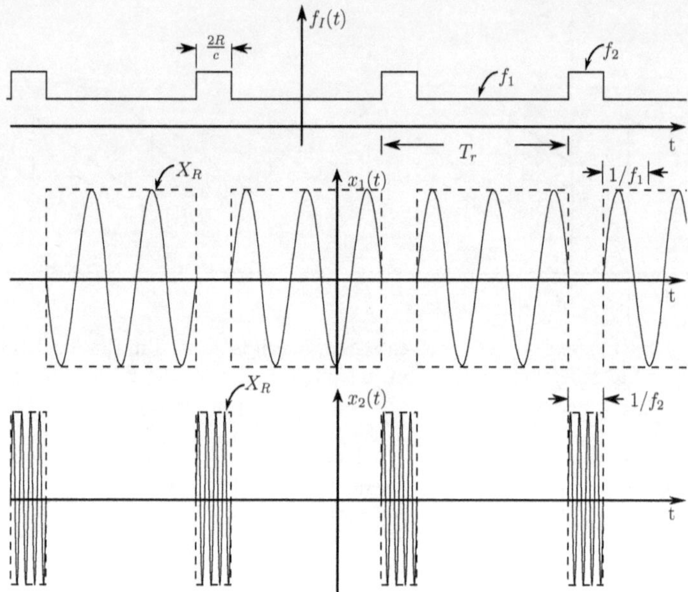

Figure 2.4: Frequency and amplitude visualization of the received signal after dechirping.

similar to Tab. 2.1. The second term in the linear factor is small with respect to the first one under consideration that for all relevant cases the target velocity is small compared to speed of light $\frac{v}{c} \ll 1$ and the Doppler-shift is negligible with respect to bandwidth B. Hence, the intermediate phase simplifies to

$$\phi_I(t_i) \approx \phi_0 - 2\pi \left[\left(\frac{2v}{c} \right) f_c + B f_r t_0 \right] t_i, \tag{2.21}$$

and accordingly, within the first pulse the frequency is

$$f_I = \underbrace{\left(\frac{2v}{c} \right) f_c}_{\text{velocity}} + \underbrace{B f_r t_0}_{\text{range}}, \tag{2.22}$$

where the first term is due to target velocity and the second term due to time delay, respectively range, to the target. Note that it is not possible to separate range and target velocity by measuring frequency f_I within one single pulse.

Consequently, the phase from the n-th pulse is examined to obtain the velocity by setting the interval $(2n - 1)T_r/2 < t < (2n + 1)T_r/2$ where n is the pulse number. We assume that the first pulse is centered at $t = 0$. The time delay t_d to the target is then

$$t_d = 2R/c = 2R_0/c + 2vt/c = t_0 + \frac{2v(nT_r + t_i)}{c}, \tag{2.23}$$

where the time to the n-th pulse is nT_r. Again, the time is substituted into the phase with

$$
\begin{aligned}
\phi_{In}(t_i, n) = & -2\pi f_c t_0 - 2\pi f_c \left(\frac{2v}{c}\right) t_i - 2\pi f_c \left(\frac{2v}{c}\right) nT_r \\
& + \pi B f_r \left[t_0 + \left(\frac{2v}{c}\right) \left(nT_r + t_i\right) \right]^2 \\
& - 2\pi B f_r \left[t_0 + \left(\frac{2v}{c}\right) \left(nT_r + t_i\right) \right],
\end{aligned}
\tag{2.24}
$$

and after expansion and elimination of small terms the phase can be expressed as

$$
\phi_{In}(t_i, n) = \phi_0 - 2\pi f_c \left(\frac{2v}{c}\right) nT_r - 2\pi \left[\left(\frac{2v}{c}\right) f_c + B f_r t_0 + \left(\frac{2v}{c}\right) Bn \right] t_i,
\tag{2.25}
$$

and the frequency f_{In} in the n-th pulse is

$$
f_{In} = \left(\frac{2v}{c}\right) f_c + B f_r t_0 + \left(\frac{2v}{c}\right) B\,n.
\tag{2.26}
$$

The frequency f_{In} in the n-th pulse is identical to that in the first pulse (Eq. 2.22) except for the third term (Eq. 2.26). The third term describes the target motion from pulse to pulse, the range $R_0 + vnT_r$, respectively.

Two effects occur within the pulse, first, its width $T = T_r - t_d$ changes slightly due to chirp dependence in t_d, and secondly, its phase changes from pulse to pulse, which is the basis for the velocity observation. Note that, the phase term $(2v/c)Bn$ in Eq. 2.25 can be neglected since it describes the phase change due to target velocity within the n-th chirp, which is small compared to the other terms. The received signal can be expressed as

$$
x_{IF}(t_i, n) = X_{IF} \cos 2\pi \left\{ \underbrace{\left[f_c \left(\frac{2v}{c}\right) + \frac{B}{T} \left(\frac{2R_0}{c}\right) \right]}_{\text{frequency present during one chirp}} t_i \right.
$$

$$
\left. + \underbrace{n f_c \left(\frac{2v}{c}\right) T_r}_{\substack{\text{chirp phase} \\ \text{change due to velocity}}} + \underbrace{f_c \left(\frac{2R_0}{c}\right)}_{\text{constant phase}} \right\}.
\tag{2.27}
$$

Accordingly, the velocity can be obtained evaluating the phase rate of change from one sweep to the subsequent. The target range can be determined from knowing the phase rate of change during one sweep. It is notable that f_c does not change greatly from sweep to sweep and that fluctuations of the constant phase term do not interfere with or mask the $n f_c (2v/c) T_r$ term. The signal enables the estimation of range and velocity for arbitrary, detectable targets.

In real-world environment perception applications, one is also interested in the relative angle. A popular method for angle estimation is direction of arrival (DoA) technique. Due to the azimuthal angle under which the target is illuminated, a path difference at the ULA receives antenna ensues and thus, a detectable phase

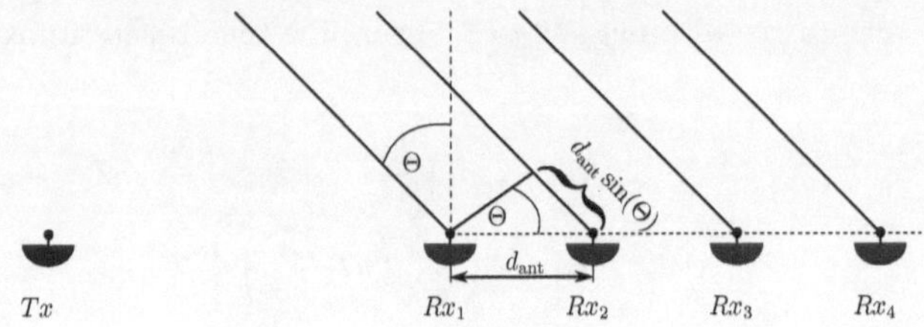

Figure 2.5: Electromagnetic wave for an exemplary incidence angle Θ.

difference occurs. An overview of DoA methods and the basis for this subsection is given in [God97].

The phase difference is caused by the target azimuth angle when the reflected electromagnetic wave is, depending on the angle, either arriving first on the left or right end of the ULA antenna array. The sampled signal across all antenna elements is linear and its phase depends on the angle of incidence. Fig. 2.5 shows the path difference of the incident wave. An exemplary ULA, consisting of four equidistant placed receiving antennas, is subject to the illumination of a far-field wavefront. The wavefront impinges first on the left antenna element Rx_1 and subsequently on elements $Rx_2...Rx_4$.

Hence, each antenna receives the signal as a superposition of sinusoidal terms for each target k with corresponding amplitude a_i and phase which can be expressed as

$$a(l) = \sum_{k=1}^{K} a_i e^{-j2\pi \frac{l \cdot d_{\mathrm{ant}}}{\lambda}\sin(\Theta_K)}. \tag{2.28}$$

where Θ_k is the azimuthal angle of target k and $l = 1, ..., L$ is the corresponding receiver antenna. The angular resolution depends on the number of receive antennas. Hence, the full three-dimensional complex radar signal of a superposition of K targets is given by

$$x_{IF}(t_i, n, l) = \sum_{k=1}^{K} \xi_k \underbrace{e^{j2\pi\left(\frac{2f_c R_0}{c}\right)}}_{\text{const. phase}} \cdot \underbrace{e^{j2\pi\left(\frac{f_c 2v}{c} + \frac{2R_k B}{cT}\right)t_i}}_{\text{range term}} \cdot$$
$$\underbrace{e^{j2\pi\left(\frac{n f_c 2v_k}{c}\right)T_r}}_{\text{velocity term}} \cdot \underbrace{e^{j2\pi\left(\frac{f_c d_{\mathrm{ant}}\sin(\Theta_k)}{c}(l-1)\right)}}_{\text{azimuth term}}. \tag{2.29}$$

The amplitude parameter ξ_k depends on path loss attenuation $1/R^4$ [Fri46], additional attenuation due to adverse weather conditions [HKDB16, HDKR17] and on target parameter like RCS. The amplitude can also be subject to multipath reflections that occur on road surfaces between the sensor and the target object, where the phase differences interfere with each other and may attenuate the received power level and degrade target detection probability [WHC16].

2.2.3 Signal processing

The down-converted electromagnetic wave captured at the receive antennas is sampled at high frequencies, e.g., 10 MHz, leading to high data rates that need to be processed. The radar signal processing aims to extract information of interest from the RF-signal in the time domain to generate a target list. Therefore, the range, velocity and azimuthal angle for every target, detected via thresholding and clustering, is determined. The pre-processed data are subject to a successive association and tracking procedure, estimating additional object parameters, e.g., target acceleration. These objects are the basis for current and future safety and automated driving functions.

For this work, the chirps from one coherent processing interval (CPI) are subject to three successively applied fast Fourier transform (FFT) on chirps, frames and arrays. Fig. 2.6 shows an exemplary radar data cube. The signal processing techniques are presented in this section and based on [CC84, Bar73]. The derivation of the continuous signal phase from Eq. 2.29 is discretized and rewritten with neglected constant phase term. Assuming constant target velocity during a single coherent measurement interval and fast time sampling with $t_i = T_s n_r$ and $T_r = T_c n_v$ to

$$\varphi_{IF}(n_r, n_v, l; r, v, \Theta) \approx \underbrace{\frac{4R\alpha}{c} T_s\, n_r}_{F_r} + \underbrace{\frac{2f_c n}{c} T_c\, n_v}_{F_v} + \underbrace{\frac{f_c d_{\mathrm{ant}} \sin(\Theta)}{c}(l-1)}_{F_\Theta}, \qquad (2.30)$$

for

$$n_r = 1, ..., N_r, \quad n_v = 1, ..., N_v, \quad l = 1, ..., L, \qquad (2.31)$$

where $\alpha = B/(2T)$, n_r, n_v and l are corresponding sample, pulse, and antenna variable, N_r, N_v, and L are the maximum number of samples per chirp, chirps per CPI and number of antenna, respectively. The corresponding frequencies F_r, F_v and F_Θ directly depend on the parameters of interest [Win14]. The radar data can be arranged in a three-dimensional radar data cube in favor of the following signal processing, see Fig 2.6. A practical method for frequency estimation is spectrum analysis using a discrete Fourier transform (DFT) considering the sample and pulse variable computed by a FFT,

$$X_{IF}(i_r, i_v, l) = \sum_{l=0}^{L-1} \sum_{n_v=0}^{N_v-1} \sum_{n_r=0}^{N_r-1} w_\Theta(l) w_v(n_v) w_r(n_r)\, x_{IF}(n_r, n_v, l)$$
$$\cdot\, e^{-j2\pi i_r n_r/N_r}\, e^{-j2\pi i_v n_v/N_v}\, e^{-j2\pi i_\Theta l/L}, \qquad (2.32)$$

for

$$i_r = 0, ..., I_r - 1, \quad i_v = 0, ..., I_v - 1, \quad i_\Theta = 0, ..., L - 1, \qquad (2.33)$$

where i_r, i_v, and i_Θ are range, velocity and azimuth angle cell indexes of the data cube, $w_r(n_r)$, $w_v(n_v)$, and $w_\Theta(l)$ are normalized window functions of lengths N_r, N_v,

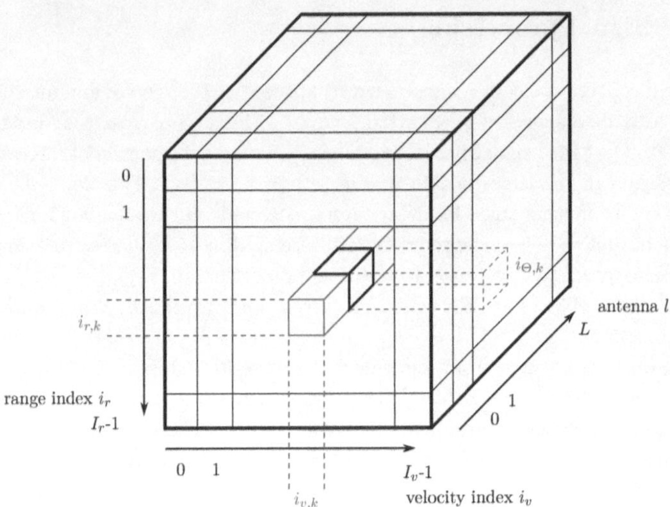

Figure 2.6: Radar data cube structure arranges range-corresponding samples from each chirp in rows (first dimension), velocity-corresponding samples in lines (second dimension) and number of antennas in the third dimension for a coherent radar measurement cycle. A snapshot at $i_{r,k}$, $i_{v,k}$, and $i_{\Theta,k}$ is indicated as solid, black cell within the full data cube.

and L respectively. Windowing is employed to suppress leakage which is a result of the side lobe structure of the spectrum due to finite data length. A possible solution applies, e.g., a Hann window, to weight the data, reduce discontinuities in the data and thereby reduce the side lobe in the frequency domain at the costs of a wider main lobe [CC84]. The DFT parameters i_r, i_v, and i_Θ relate the normalized frequency values to the parameter of interest by

$$i_r \simeq F_r N_r = \frac{4\alpha R}{c} T_s N_r, \tag{2.34}$$

$$i_v \simeq F_v N_v = \frac{2 f_c v n}{c} T_c N_v, \tag{2.35}$$

$$i_\Theta \simeq F_\Theta (L-1) = \frac{f_c d_{\mathrm{ant}} \sin(\theta)}{c}(L-1). \tag{2.36}$$

The down-converted, band-pass filtered and sampled *IF*-signal measurements are real-valued and can therefore be considered as $I_r < N_r/2$. The values of the data cube cells are complex-valued after application of the first Fourier transformation, so that the Doppler spectrum can be considered for all variables $I_v = N_v$. The cell widths, or resolution for each dimension, can be expressed with κ as wavenumber as the inverse cell parameter of interest

$$\triangle_r = \frac{c}{4\alpha T_s N_r}, \quad \triangle_v = \frac{c}{2 f_c n T_c N_v}, \quad \triangle_\Theta = \arcsin\left(\frac{1}{\kappa L d_{\mathrm{ant}}}\right). \tag{2.37}$$

2.2.4 CFAR thresholding and clustering

After frequency estimation using a 3D-FFT, the resulting periodogram is subject to a constant false alarm rate (CFAR) method, which is used for power detection within the data cube and to encounter time-varying noise and interference statistics. Generally, CFAR techniques compare the received signal amplitude to a threshold. Usually, an adaptive threshold is applied to reflect the local clutter situation due to the reflections from, e.g., rain, small objects, random noise from ground reflections or general sensor noise. The intention is to keep the false alarm probability over the entire measurement space constant and still detect all present targets.

Established CFAR methods are cell averaging constant false alarm rate (CA-CFAR), cell averaging with greatest of constant false alarm rate (CAGO-CFAR), and ordered statistic constant false alarm rate (OS-CFAR). CFAR systems usually use the sliding window technique. The first two methods use a split neighborhood, e.g., a certain number of range cells, where the arithmetic mean of the amplitude is obtained. Therefore, the first step is to determine the mean clutter power level Z for surrounding cells followed by multiplication with a scaling factor T, resulting in a threshold value TZ. After comparing the actual cell to this threshold, the cell is either a detection, when the cell value is above the threshold, or no detection when the cell value is below the threshold.

The main difference between CA-CFAR and CAGO-CFAR is that the former is based on the assumption of a uniform clutter situation in the entire neighborhood area, whereas the latter makes allowance for clutter edges occurring within the reference area. The performance evaluation in [Roh83] considers two clutter backgrounds: uniform (stationary) and non-uniform clutter within the reference window. In dense target situations, e.g., where two targets are closely spaced in range and azimuth, the echos of both may be within one reference window leading to an interpretation of one target as clutter power and hence, leaves one or even both targets undetected.

To overcome the assumption of an uniform statistic in the reference window, Rohling *et al.* proposed the ordered statistic CFAR (OS-CFAR) method, yielding superior results in contrast to CA-CFAR processing [Roh83, SV00]. The method performs a rank-ordering of the values encountered in the neighborhood area according to their magnitude. The central idea of an OS-CFAR is to select one certain value $X_k, k \in 1, 2, ..., N$ from the rank-ordered amplitude sequence in the reference window $X_1 \leq X_2 \leq ... \leq X_N$, where the indices indicate the rank-order number. X_1 denotes the minimum and X_N the maximum value. Subsequently, an estimate Z for the average clutter power observed in the reference window is generated.

The advantage of OS-CFAR method is if one or multiple targets are present, the threshold value is limited by a predefined number of neighboring cells. Therefore the number of detections per present target is higher than with a CA-CFAR method enabling a high-resolution radar image. In this work, the OS-CFAR method is used to detect targets in all three dimensions: range, velocity and azimuth angle.

After applying the OS-CFAR procedure on the radar data cube, a clustering pro-

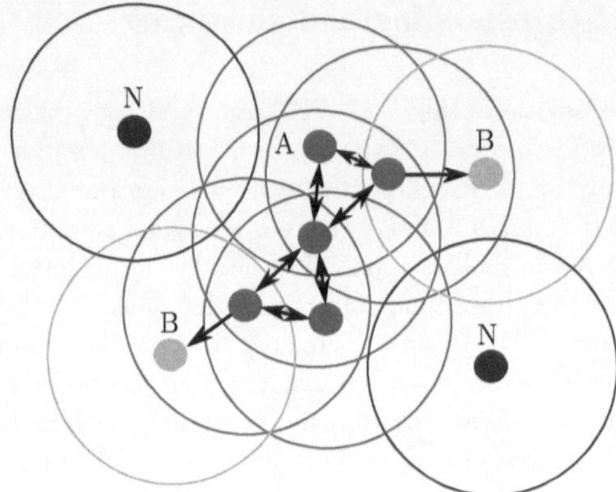

Figure 2.7: Exemplary data points, clustered using DBSCAN algorithm and Euclidean distance. Results are indicated by colors: core points (red), border points (orange), noise points (blue).

cedure is applied to cluster detections originating from one extended target, e.g., a vehicle. For subsequent object parameter estimation and tracking, merely the detections which can be assigned to one object are of interest. Therefore a well-known and robust clustering procedure density-based spatial clustering of applications with noise (DBSCAN), developed from Ester *et al.* [EKS+96], is applied and is briefly introduced.

In the automotive context, clustering algorithms are usually used to group data obtained using the sensor for class identification. The radar data, measured from arbitrarily shaped objects, depend mainly on sensor-to-object configuration, range, reflection surface and material. Hence, the multi-dimensional clusters may be arbitrary shaped and the domain knowledge shall be minimal, e.g., the number clusters in the data is unknown.

Established clustering techniques rely on similarity, e.g., in the same cluster, which shall be as similar as possible. On the other hand different clusters shall be as different as possible. Clustering algorithms can be divided into 9 categories defined in the parameter they utilize [XT15]. Popular algorithms are, e.g., based on a partition, where they construct a partition of a database D of n objects into a set of k clusters as k is an input parameter for these algorithms. Hence, some domain knowledge is required, which is usually not available for automotive applications [EKS+96].

Another conventional clustering algorithm is based on hierarchy, which aims to create a hierarchical decomposition of D. The hierarchical decomposition is represented by a dendrogram, which can be interpreted as a tree that iteratively splits D into smaller subsets until each subset consists of a single object. In such a hierarchy, each node of the tree represents a cluster of the database D. Contrary to partitioning algorithms, hierarchical algorithms do not need the number of clusters k as input. Instead, the demand for a termination condition, which has to be defined indicat-

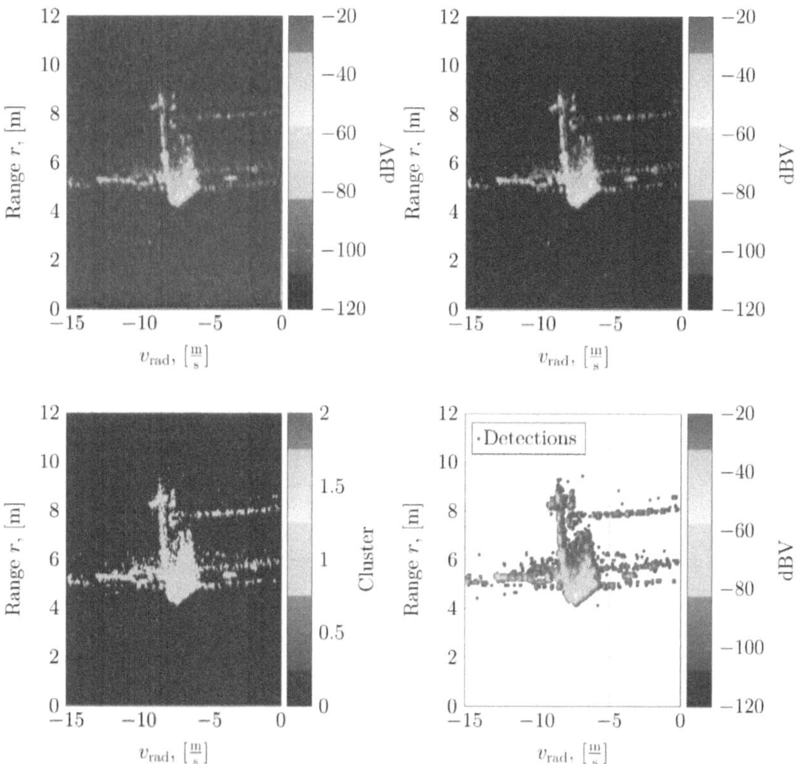

Figure 2.8: Exemplary real radar data visualization of the presented raw data signal processing including FFT, CFAR, DBSCAN and detection list representation for an approaching vehicle.
Top left: Raw range-Doppler snapshot after FFT application. **Top right**: Range-Doppler snapshot after CFAR filtering. **Bottom left**: Range-Doppler snapshot after DBSCAN. **Bottom right**: Final range-Doppler snapshot data as detection list. Each data point represents a detection exceeding the CFAR-threshold and is assigned to the extended object.

ing when the merge or division process should be terminated, e.g., when a minimal critical distance D_{min} between clusters is deceeded.

A third clustering approach is based on the density between data points. The key idea of density-based clustering is that for each point of a cluster, the neighborhood of a given radius has to exceed some threshold number of points. The shape of a neighborhood is determined by choice of an adequate distance function, where this approach works with any distance function so that an appropriate function can be chosen for some given application [EKS+96].

The DBSCAN procedure groups detections based on the maximum distance parameter and the minimum number of data points within this distance (tuning parameters), into three categories: core points, border points, noise points. Fig. 2.7 shows an exemplary data set and the result of the DBSCAN. The initial point is arbitrarily selected, e.g., A, and retrieves all points density-reachable from this point concerning the two tuning parameters. This point is set as a core point if the conditions using the tuning parameters are fulfilled and a cluster is yielded, whereas if the point is a border point, e.g., B, no points are density-reachable from this point and the next data point is analyzed. If there is not a single point within the distance tuning parameter, then this point is tagged as a noise point.

In this work, the DBSCAN algorithm is used to identify and group dense radar detections, originating from one object. Since target vehicles are assumed to be moving within this work, the DBSCAN procedure is applied in the range-velocity dimensions of the radar data. Additional or extended DBSCAN iterations considering the angle information may be beneficial to the clustering result.

2.2.5 Real radar data example

Fig. 2.8 shows the signal processing chain starting with an 2D-FFT to process one data frame. The top left plot in Fig. 2.8 presents the 2D-FFT result showing the target vehicle with various velocity components. The top right plot in Fig. 2.8 shows that the signal-to-noise ratio (SNR) of the target is increased due to OS-CFAR procedure application compared to the previous processing step. The clustering of detections that originate from the target is archived using the DBSCAN procedure. The result is shown in the bottom left plot in Fig. 2.8. The clustered detection list and final result is shown in the bottom right plot in Fig. 2.8 as a high-resolution detection list of the approaching vehicle.

Overall, an enhanced target detection performance may be achieved if the data processing steps FFT, CFAR and DBSCAN are applied as in the proposed sequence. Variations and multiple applications of one or more processing steps may increase the target detection performance, and, on the other hand, may decrease algorithm run-time. However, for this work the proposed raw data signal processing procedure is emphasized in terms of computational complexity, algorithm run-time and target detection performance.

2.3 Summary

This chapter presents the fundamental electromagnetic wave propagation in free space and introduces the FMCW radar system, which is used in multiple studies within this work. The FMCW radar operates using the chirp sequence operation mode where fast subsequent chirps are sent, received, down-converted to the baseband and processed. The chapter provides a comprehensive derivation of the FMCW chirp sequence operation mode. The information of targets, obstacles and free spaces is available after a spectral analysis, e.g., using a FFT. Additional signal processing effort is made using thresholding (OS-CFAR) and clustering procedure (DBSCAN) to enhance target detection performance and to assign detections to one target object. A comparison of the proposed procedures is given. At last, the proposed signal processing procedure proves sufficient target detection performance by an exemplary application on a real radar data snapshot for an approaching vehicle on a collision trajectory.

References

[Bar73]   Donald E. Barrick. FM/CW radar signals and digital processing. Technical report, National oceanic and atmospheric administration, 1973.

[CC84]   Edmond A. Costa and Russell B. Chadwick. Signal processing and range spreading in the FM-CW radar. Technical report, National oceanic and atmospheric administration, 1984.

[CL82]   Kent Chamberlin and Raymond Luebbers. An evaluation of Longley-Rice and GTD propagation models. *IEEE Transactions on Antennas and Propagation*, 30(6):1093–1098, 1982.

[EKS$^+$96]   Martin Ester, Hans-Peter Kriegel, Jörg Sander, Xiaowei Xu, et al. A density-based algorithm for discovering clusters in large spatial databases with noise. In *Kdd*, volume 96, pages 226–231, 1996.

[Fon55]   Karl Fonck. Radar bremst bei Gefahr. *Auto, Motor & Sport*, 22:30, 1955.

[Fri46]   Harald T. Friis. A note on a simple transmission formula. *proc. IRE*, 34(5):254–256, 1946.

[God97]   Lal C. Godara. Application of antenna arrays to mobile communications. II. Beam-forming and direction-of-arrival considerations. *Proceedings of the IEEE*, 85(8):1195–1245, 1997.

[Had19]   Andreas Haderer. RDL-77G-TX2RX16 Frontend (User Manual). Technical report, INRAS, 2019.

[HAS08]   James M. Headrick, Stuart J. Anderson, and Merrill Skolnik. HF over-the-horizon radar. *Radar handbook*, 3, 2008.

[HDKR17]   Sinan Hasirlioglu, Igor Doric, Alexander Kamann, and Andreas Riener. Reproducible fog simulation for testing automotive surround sensors. In *2017 IEEE 85th Vehicular Technology Conference (VTC Spring)*, pages 1–7. IEEE, 2017.

[IIKDB16]   Sinan Hasirlioglu, Alexander Kamann, Igor Doric, and Thomas Brandmeier. Test methodology for rain influence on automotive surround sensors. In *2016 IEEE 19th International Conference on Intelligent Transportation Systems (ITSC)*, pages 2242–2247. IEEE, 2016.

[JJL14]   Yeonghwan Ju, Youngseok Jin, and Jonghun Lee. Design and implementation of a 24 GHz FMCW radar system for automotive applications. In *2014 International Radar Conference*, pages 1–4. IEEE, 2014.

[KHP+18] Alexander Kamann, Patrick Held, Florian Perras, Patrick Zaumseil, Thomas Brandmeier, and Ulrich T. Schwarz. Automotive radar multipath propagation in uncertain environments. In *2018 21st International Conference on Intelligent Transportation Systems (ITSC)*, pages 859–864. IEEE, 2018.

[Rem16] Bernhard Rembold. *Wellenausbreitung: Grundlagen–Modelle–Messtechnik–Verfahren*. Springer-Verlag, 2016.

[Roh83] Hermann Rohling. Radar CFAR thresholding in clutter and multiple target situations. *IEEE transactions on aerospace and electronic systems*, pages 608–621, 1983.

[SV00] Michael E. Smith and Pramod K. Varshney. Intelligent CFAR processor based on data variability. *IEEE Transactions on Aerospace and Electronic Systems*, 36(3):837–847, 2000.

[WHC16] Hsiao-Ning Wang, Ying-Wei Huang, and Shyh-Jong Chung. Spatial diversity 24-GHz FMCW radar with ground effect compensation for automotive applications. *IEEE Transactions on Vehicular Technology*, 66(2):965–973, 2016.

[WHW09] Hermann Winner, Stephan Hakuli, and Gabriele Wolf. *Handbuch Fahrerassistenzsysteme: Grundlagen, Komponenten und Systeme für aktive Sicherheit und Komfort: mit 550 Abbildungen und 45 Tabellen*. Springer-Verlag, 2009.

[Win14] Markus Wintermantel. Radar system with improved angle formation, March 4 2014. US Patent 8,665,137.

[XT15] Dongkuan Xu and Yingjie Tian. A comprehensive survey of clustering algorithms. *Annals of Data Science*, 2(2):165–193, 2015.

Chapter 3

Target detection

This chapter presents the principles of electromagnetic wave properties in the near- and far-field, corresponding Fresnel and Fraunhofer region. In close sensor-to-target configurations, the micro-Doppler effect of vibrating or rotating target components, e.g., spinning wheels, is detectable and provides additional, valuable target information. The micro-Doppler effect for spinning wheels is derived and real radar data snapshots present the measured micro-Doppler effect with a high-resolution radar sensor. These Doppler signals are subject to a spinning wheel position and corresponding bulk velocity estimation method based on spatially resolved micro-Doppler spectra. This novel method is presented and evaluated in a high dynamic evading maneuver using a test vehicle as target. The presented method serve as input for the subsequent chapter on object tracking. The wheel detection method copes with the wandering dominant scatter point of the target vehicle, as described in the original contributions section 1.3.

In comparison to the directly backscattered signals from the target object back to the sensor, target vehicles in close distance may be additionally subject to multipath propagated signals. The transmitted electromagnetic wave potentially gets reflected by other objects or structure surfaces like, e.g., guardrails as described in the original contributions section 1.3. To address this problem, this chapter introduces a method to identify potential false-positive detections based on multipath propagation.

3.1   Background

This section introduces the Fresnel and Fraunhofer region and describes its fundamental criterions. The summary given is based on the work of Mahafza *et al.* [Mah05] for phased array antennas. Besides, the section introduces the micro-Doppler effect, which is generated by vibrating or rotating parts of the moving target, e.g., target vehicle wheels.

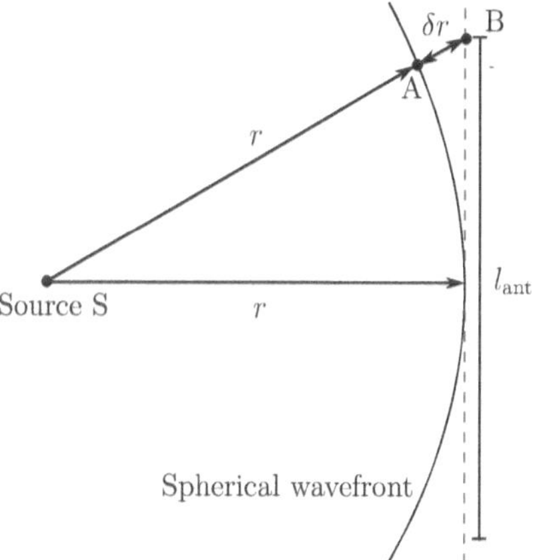

Figure 3.1: Spherical wavefront geometry to derive the far-field criterion.

3.1.1 Fresnel and Fraunhofer region

Depending on the distance from the transmitting antennas to the target, three distinct regions are identified: the near-field, Fresnel and Fraunhofer regions. In the near-field and Fresnel regions, the emitted rays have spherical wavefronts that correspond to equiphase wavefronts, whereas in the Fraunhofer region, the wavefront can be locally represented by plane wavefronts. Usually, near-field and Fresnel regions are of little interest for most radar applications, whereas within this **book**, the Fresnel region is of interest due to the close sensor-to-target distance in pre-crash scenarios. Most radar applications operate in the Fraunhofer region, which is also referred to as the far-field region.

Fig. 3.1 show a spherical wavefront to derive the far-field criterion. The criterion determines the minimal distance for given antenna apertures where plane wavefronts form. Therefore, one consider a radiating source S that emits spherical waves. A receiving antenna of length l_{ant} is at distance r from the radiation source. The phase difference between the spherical wave and a local plane wave at the receiving antenna can be expressed as the distance δr, which is given by [Mah05]

$$\delta r = \overline{SA} - \overline{SB} = \sqrt{r^2 + \left(\frac{l_{\mathrm{ant}}}{2}\right)^2} - r, \tag{3.1}$$

and due to the fact that $r \gg l_{\mathrm{ant}}$ in the far-field the expression can be approximated via binomial expansion by

$$\delta r = r\left(\sqrt{1 + \left(\frac{l_{\mathrm{ant}}}{2r}\right)^2} - 1\right) \approx \frac{l_{\mathrm{ant}}^2}{8r}. \tag{3.2}$$

It is conventional to assume plane waves when the distance δr corresponds to less than 1/16 of a wavelength or more precisely

$$\delta r = \frac{l_{\text{ant}}^2}{8r} \leq \lambda/16. \tag{3.3}$$

A expression for the far-field criterion is then

$$r \geq \frac{2l_{\text{ant}}^2}{\lambda}, \tag{3.4}$$

where it is notably that the far-field function depends on both the antenna size and operating wavelength. For most measurements in this **book** the INRAS radar log with an operating frequency of 77 GHz and an receiver patch antenna aperture of approximately 0.15 m is used. According to the far-field criterion the Fresnel region ends and Fraunhofer region starts at 12 m.

3.1.2 Target detection properties

The previous chapter introduced the radar raw data signal processing up to a list of detections per present object. For the sake of clarification, a brief definition of target assumptions is given in this section. The high radar resolution in range, velocity, and azimuthal angle may lead to multiple detections per object. More than one detection per object yield additional information about the target, which is of major interest in this **book**. The different target definitions are based on [Kel17]. The number of possible detections per object depend on the object geometry with respect to the sensor resolution. Every range, velocity, and azimuth bin may contain one detection. If the physical dimension of the object is smaller than the resolution in the three dimensions, the object causes one detection and can be regarded as *point target*. Point targets are the simplest target representation type.

If the target dimensions exceed either range, velocity, and azimuthal resolution and detections of the same object are distributed over multiple neighboring detection bins, the target can be referred to as *extended object*. Extended target state estimation, based on multiple detections of, e.g., radar or lidar data, is of major interest since the sensor resolution is continuously increasing, e.g., [HLS12, HSSS12]. An extended target may cause only one detection and therefore change from extended target to a point target, e.g., due to increasing distance between sensor and target. In contrast, the target may have a small size with respect to the sensor resolution and, therefore, only possesses an extend in the velocity dimension. The target is referred to as *kinematic extended object*, e.g., pedestrians in specific sensor-to-target configurations. The kinematic extension is induced either by the micro-Doppler effect or due to multiple radial velocities in the equivalent range and angle bin.

In most of the measurements and test runs carried out in this **book**, the targets were either real vehicles or retroreflectors, extended and point targets, respectively. A retroreflector consists of three mutually perpendicular, intersecting flat surfaces, which reflects waves directly back towards the source.

3.1.3 Micro-Doppler effect

Besides the number of detections, another physical effect is of major interest within this **book** for close target detection – the *micro-Doppler* effect. The micro-Doppler effect was originally introduced in the context of incoherent laser systems [ZRH98]. According Chen *et al.* [CLHW06], the micro-Doppler effect occurs when the target or any structure on the target has mechanical vibrations or rotation in addition to its bulk translation. An additional frequency modulation may be induced on the returned signal that generates sidebands about the target Doppler frequency shift [CLHW03, PLH92]. This includes returned radar signals such as propellers of a fixed-wing aircraft and rotating wheels of a vehicle. Micro-Doppler signals may serve as additional target features. The fundamental micro-Doppler analysis was presented by Chen *et al.* [CLHW03, CLHW06], where the authors analyzed micro-Doppler signatures of vibrating or rotating structures, modeled as point scatterers, in the time-frequency domain using computer simulations. A periodic vibration or rotation generates sideband Doppler frequency shifts about the Doppler shifted central carrier frequency. The modulation involves harmonic frequencies that depend on the carrier frequency, the vibration or rotation rate, the angle between the direction of vibration and the direction of the incident wave.

According to Chen *et al.*, the micro-Doppler effect can be derived by introducing vibration or rotation to conventional Doppler analysis. Fig. 3.2 shows an exemplary spinning wheel that is moving away from a stationary radar sensor. The micro-Doppler is derived for this spinning wheel example considering a point scatterer P as simplification. Note that the number of point scatterers may be increased arbitrarily. The radar system is assumed stationary and located at the origin Q. Three coordinate systems are introduced, the radar coordinate system (U, V, W), the target coordinate system attached to the target (x, y, z) and a reference coordinate system (X, Y, Z) which has the same translation as the target local coordinates (x, y, z) but has no rotation with respect to the radar coordinates (U, V, W). The reference coordinate system shares the identical origin O with the target local coordinates and is assumed to be at a distance R_0 from the radar. The unit vector of the radar LoS is defined as

$$ \boldsymbol{n} = \frac{\boldsymbol{R_0}}{\|\boldsymbol{R_0}\|} = (\cos(\alpha)\cos(\beta), \sin(\alpha)\cos(\beta), \sin(\beta))^T, \qquad (3.5) $$

where $\| \cdot \|$ is the Euclidean norm, α the azimuth and β the elevation angle of the target in the radar coordinates (U, V, W), respectively.

The target has a translational velocity $\boldsymbol{v}_{\text{trans}}$ and an angular rotation velocity $\boldsymbol{\omega}$ with respect to the radar, where the angular rotation velocity can be represented in the reference coordinate system as $\boldsymbol{\omega} = (\omega_X, \omega_Y, \omega_Z)^T$. The point scatter P, which is initially located at $\boldsymbol{r}_0 = (X_0, Y_0, Z_0)^T$ is assumed to move from its initial location to point P' within time t and velocity $\boldsymbol{v}_{\text{trans}}$. The motion can be decomposed in a translational and rotational motion and the distance from sensor to P' can be

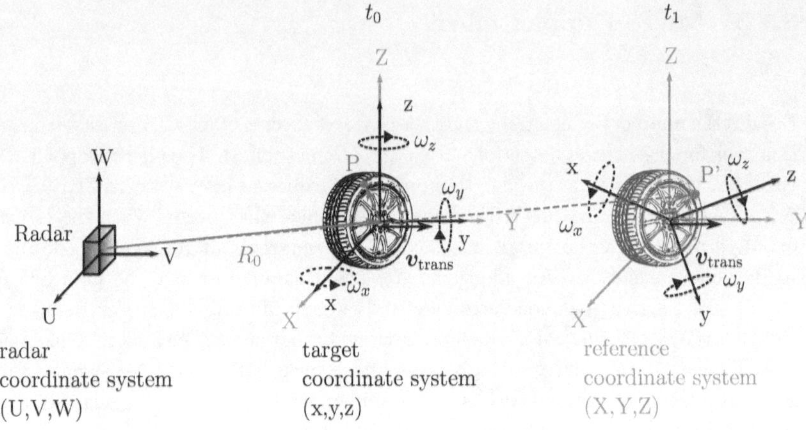

Figure 3.2: Geometry of a radar sensor which measures a spinning wheel moving away from the sensor and thus, exhibiting a translational and rotational velocity component. The various coordinate systems and point P is used to derive the micro-Doppler equations.

expressed as

$$\overline{QP'} = \boldsymbol{R_0} + \boldsymbol{v}_{\text{trans}}t + \text{Rot}(t) \cdot \boldsymbol{r_0}, \tag{3.6}$$

where the rotational component is described by $\text{Rot}(\bullet)$, R_0 is the distance from the radar to the initial point O, $\boldsymbol{v}_{\text{trans}}$ is the transitional velocity component within time interval t and $\boldsymbol{r_0}$ is the distance from O to scatter point P. Note that, the rotational motion at each time interval t is assumed infinitesimal. The time-dependent scalar range measured in the sensor coordinate system is accordingly

$$R(t) = \|\boldsymbol{R_0} + \boldsymbol{v}_{\text{trans}}t + \text{Rot}(t) \cdot \boldsymbol{r_0}\|. \tag{3.7}$$

The radar transmits an exemplary sinusoidal waveform with a carrier frequency f_c and the returned baseband signal from the point scatterer is a function of $R(t)$ as

$$s(t) = \rho(x,y,z) \cdot e^{\left(j2\pi f_c \frac{2R(t)}{c}\right)} = \rho(x,y,z) \cdot e^{(j\Phi_{\mu D}(R(t)))}, \tag{3.8}$$

where $\rho(x,y,z)$ is the point scatterer P reflectivity function in local coordinates (x,y,z). The baseband signal is

$$\Phi_{\mu D}(R(t)) = 2\pi f_c \frac{2R(t)}{c}, \tag{3.9}$$

and after deriving the phase the Doppler frequency shift of the target motion can be expressed as

$$
\begin{aligned}
f_D &= \frac{1}{2\pi}\frac{d\Phi_{\mu\mathrm{D}}(R(t))}{dt} = \frac{2f}{c}\frac{d}{dt}R(t), \\
&= \frac{2f}{c}\frac{1}{2R(t)}\frac{d}{dt}\left[(\boldsymbol{R_0} + \boldsymbol{v}_{\mathrm{trans}}t + \mathrm{Rot}(t)\cdot\boldsymbol{r_0})^T \times (\boldsymbol{R_0} + \boldsymbol{v}_{\mathrm{trans}}t + \mathrm{Rot}(t)\cdot\boldsymbol{r_0})\right], \\
&= \frac{2f}{c}\left[\boldsymbol{v}_{\mathrm{trans}} + \frac{d}{dt}(\mathrm{Rot}(t)\cdot\boldsymbol{r_0})\right]^T \boldsymbol{n}_P,
\end{aligned}
\tag{3.10}
$$

where

$$
\boldsymbol{n}_P = \frac{\boldsymbol{R_0} + \boldsymbol{v}_{\mathrm{trans}}t + \mathrm{Rot}(t)\cdot\boldsymbol{r_0}}{\|\boldsymbol{R_0} + \boldsymbol{v}_{\mathrm{trans}}t + \mathrm{Rot}(t)\cdot\boldsymbol{r_0}\|},
\tag{3.11}
$$

is the direction unit vector from the radar to P'. The rotational component of Eq. 3.10 can be expressed as a skew symmetric matrix $\hat{\boldsymbol{\omega}}$ associated with $\boldsymbol{\omega}$, which is the linear transformation that computes the cross product of the vector $\boldsymbol{\omega}$ with any other vector. Therefore, the angular rotation velocity vector $\boldsymbol{\omega} = (\omega_X, \omega_Y, \omega_Z)^T$ rotates along the unit rotation vector $\boldsymbol{\omega}' = \boldsymbol{\omega}/\|\boldsymbol{\omega}\|$. Assuming that, the rotational motion at each time interval is infinitesimal the rotation matrix can be written in terms of matrix $\hat{\boldsymbol{\omega}}$ as

$$
\mathrm{Rot}(t) = e^{\hat{\boldsymbol{\omega}}t},
\tag{3.12}
$$

with

$$
\hat{\boldsymbol{\omega}} = \begin{bmatrix} 0 & -\omega_Z & \omega_Y \\ \omega_Z & 0 & -\omega_X \\ -\omega_Y & \omega_X & 0 \end{bmatrix}.
\tag{3.13}
$$

Thus, the Doppler frequency shift can be expressed as

$$
\begin{aligned}
f_D &= \frac{2f}{c}\left[\boldsymbol{v}_{\mathrm{trans}} + \frac{d}{dt}(e^{\hat{\boldsymbol{\omega}}t}\boldsymbol{r_0})\right]^T \boldsymbol{n}_P, \\
&= \frac{2f}{c}\left(\boldsymbol{v}_{\mathrm{trans}} + \hat{\boldsymbol{\omega}}e^{\hat{\boldsymbol{\omega}}t}\right)^T \boldsymbol{n}_P, \\
&= \frac{2f}{c}\left(\boldsymbol{v}_{\mathrm{trans}} + \hat{\boldsymbol{\omega}}\boldsymbol{r}\right)^T \boldsymbol{n}_P, \\
&\approx \frac{2f}{c}(\boldsymbol{v}_{\mathrm{trans}} + \hat{\boldsymbol{\omega}} \times \boldsymbol{r})^T \boldsymbol{n},
\end{aligned}
\tag{3.14}
$$

where the direction unit vector $\boldsymbol{n}_P$ can be approximated by $\boldsymbol{n} = \boldsymbol{R_0}/\|\boldsymbol{R_0}\| \approx \boldsymbol{n}_P$ due to $\boldsymbol{R_0} \gg \|\boldsymbol{v}_{\mathrm{trans}}t + \mathrm{Rot}(t)\cdot\boldsymbol{r}\|$. Hence, the Doppler frequency shift is approximately

$$
f_D = \frac{2f}{c}\left[\boldsymbol{v}_{\mathrm{trans}} + \boldsymbol{\omega} \times \boldsymbol{r}\right]_{\mathrm{radial}},
\tag{3.15}
$$

where the first term is the Doppler shift due to the translational motion and the second term is the term for the micro-Doppler shift caused by the rotation of the object

$$f_{\text{micro-Doppler}} = \frac{2f}{c} \left[\boldsymbol{\omega} \times \boldsymbol{r} \right]_{\text{radial}} .$$
(3.16)

Beyond Chen, literature reports some approaches which deal with the extraction and valid interpretation of the micro-Doppler effect for traffic participants in motion. A first approach to detect spinning wheels from a moving target under utilization of the micro-Doppler effect for automotive applications was presented by Kellner *et al.* [KBK+15, DKH+15]. In this approach, the authors introduce a normalized Doppler moment of each reflection, which describes the Doppler signature based on the wheel Doppler distributions. A bulk motion corridor is formed and the received detections outside this corridor are weighted according to the velocity difference to the bulk motion. The wheel-induced micro-Doppler detections are separated from the bulk motion taking the azimuthal parameter for each detection into account. This method achieved an average detection rate of 1.5 wheels per measurement. Due to the bulk corridor, it is sensitive to the specific bulk velocities.
A similar method to separate the micro-Doppler from the bulk motion Doppler using a normalization approach was presented by Li *et al.* [LDL11]. The key idea is to estimate the returned signal based on physical models. They assume a deterministic amount of scatter points on the wheel surface, and, based on these points, the authors derived a model to estimate the returned radar signal strength, which is normalized consequently.

In this **book**, the micro-Doppler effect is extensively evaluated for targets in close distances. The method delivers characteristic wheel hypotheses on the vehicle surface in a generic approach without the specification of a bulk motion velocity. The high Doppler resolution of the radar enables a comprehensive analysis of the target vehicle specific Doppler and micro-Doppler signals.

3.2 Spatially resolved micro-Doppler spectra based wheel detection

The presented FMCW radar platform and signal processing (section 2.2.1 and 2.2.3) provide range, velocity and azimuthal information of targets in the local environment. An approaching target in close distance generates detections in several range, velocity and angular bins due to the radar system's high resolution. The rotating components, e.g., target vehicle wheels, induce additional micro-Doppler signals. The literature research in the previous chapter stated a few methods and procedures to process the micro-Doppler signal and to derive additional information about the target object.

However, a generic and robust method, which requires sparse assumptions about the target, is still left to be researched. This section presents a generic method to process micro-Doppler information and provide a solution to the wandering dominant scatter point problem as one of the scientific contributions of this book.

3.2.1 Method

The key idea is to estimate spatial wheel positions in the x-y-plane and corresponding bulk velocities of the rotating components using the micro-Doppler effect to generate characteristic points fixed in the target coordinate system: the target vehicle wheels. These characteristic points are unlikely to move on the target surface. Instead, they maintain on their positions even in highly dynamic driving situations like evading maneuvers and are referred to as clustered wheel detections, which are processed to wheel hypotheses $W_{l_w}(x, y, v_B)$. After the reader is introduced to real radar data visualizing the micro-Doppler effect of an approaching target vehicle, the procedure flow chart presents the proposed method. The proposed method incorporates all pre-processed detections assigned to one target $(N_{D,\text{target}}(r, v_{\text{rad}}, \theta))$ per radar snapshot. This includes reflections from the vehicle body as well as from the rotating wheels, referred to as *Doppler detections* and *micro-Doppler detections* in the r-v_{rad}-plane, respectively.

Fig. 3.3 presents radar data snapshots, where a target vehicle approaches the sensor in different driving situations with varying criticality for each scenario: straight target approach, close target turning maneuver and evading maneuver. The turning maneuver is executed at low speed, whereas the evading maneuver is performed with a target velocity of approximately $9\,\frac{\text{m}}{\text{s}}$. Each row of Fig. 3.3 shows one of the scenarios mentioned above, where the left column presents the target detections in x-y-plane and the right column presents the Doppler- and micro-Doppler effect based detections as the velocity profile in r-v_{rad}-plane.

The first row of Fig. 3.3 presents the straight approaching maneuver. The target vehicle is at roughly 9 m. The dominant scatter point is centered on the front bumper of the vehicle and the bulk target velocity during the maneuver is at roughly $-8\,\frac{\text{m}}{\text{s}}$. The bulk induced detections are narrowly spread along the r-axis since no yaw motion of the target is present. Intense and along the v_{rad}-axis widespread front wheels

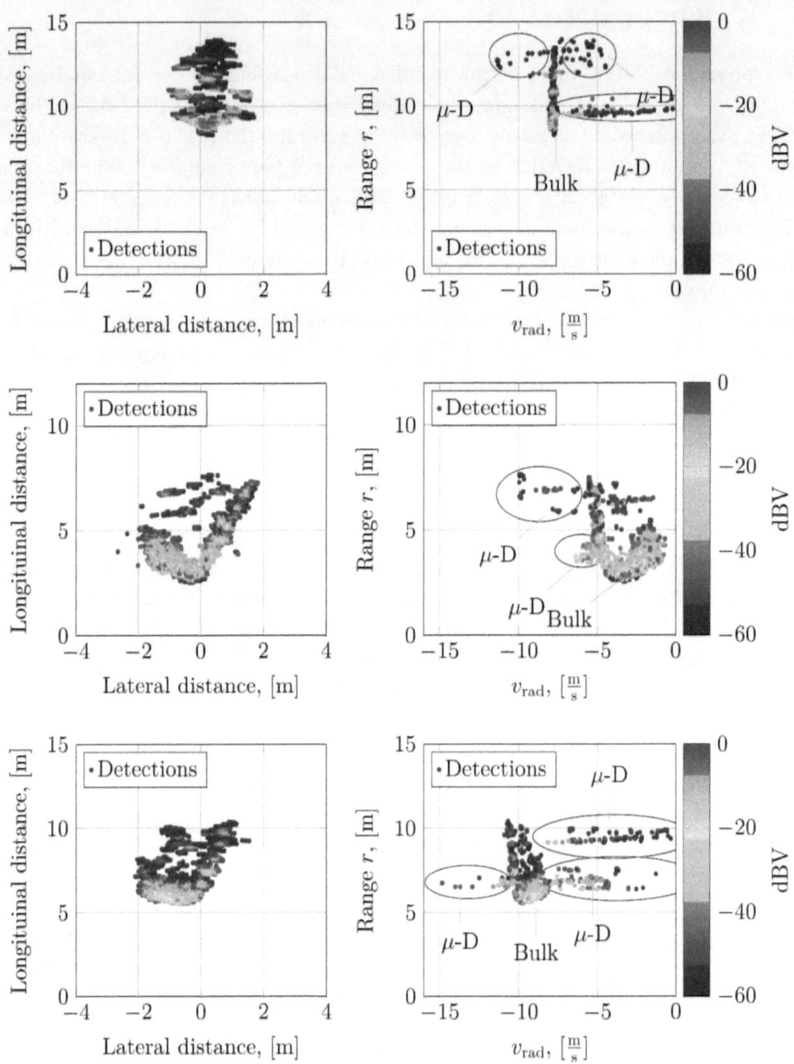

Figure 3.3: Snapshots of radar measurements presenting the target position (left column) and the target velocity profile (right column) in three target approaching scenarios: straight target approach, close target turning maneuver and critical evading maneuver. Detections mainly generated by the bulk are referred to as *Doppler detections* and detections mainly generated by rotating wheels are referred to as *micro-Doppler detections*, respectively. The bulk- and wheel-induced detections yield distinct velocity profiles (distributions along the r-v_{rad}-axis at bulk and wheel positions) for each snapshot.

induced, micro-Doppler based detections are present. Moreover, detections accumulate at back axle range, which are dispersed but clearly visible.

The second row of Fig. 3.3 presents a radar snapshot taken during a close turning maneuver in front of the radar sensor. The left front bumper is at approximately 2.2 m. Compared to the first example, the closer vehicle shows a broad dominant scatter point covering the front bumper and partially left side of the vehicle. Overall, the majority of detections originate from the target surface. The bulk velocity in the profile shows a similarly broad distribution along the v_{rad}-axis due to the close distance and high target yaw rate of the turning maneuver. The dominant scatter point velocity is approximately $-3\,\frac{\text{m}}{\text{s}}$. Notable are intense micro-Doppler detections induced by the front wheels at $-6\,\frac{\text{m}}{\text{s}}$ whereas the back axle wheels induce a widespread group of detections along the r-axis.

The bottom row of Fig. 3.3 shows a snapshot during an evading maneuver. The left front bumper is at roughly 5.5 m while the vehicle maintains a high yaw rate to avoid a collision with the radar sensor. The distribution of spatial detections is similar to the turning maneuver. The target velocity profile shows again a wide bulk detections spreading along the v_{rad}-axis. The dominant scatter point velocity is $-9\,\frac{\text{m}}{\text{s}}$ in the front of the vehicle. The various micro-Doppler detections induced by the front and rear wheels are clearly detectable and visible.

These measurements proof, the wheel induced micro-Doppler signals depend strongly on the sensor-to-target configurations and are detectable with the proposed sensor. Low-range and dynamic driving situations, like pre-crash evading maneuvers, induce multiple micro-Doppler detections due to high front-wheel steering angle and high yaw rates. Besides, in pre-crash scenarios, the number of objects in the sensor FoV is radically reduced compared to other environment perception functions, e.g., for comfort systems. This leads to less interference of multiple objects in the r-v_{rad} dimension.

In the following, a procedure to estimate wheel position and corresponding bulk velocity estimation, based on the micro-Doppler signals, is presented. Fig. 3.4 presents the procedure flow chart. The key idea is a two-staged method to determine and incorporate the variance of detections along the v_{rad}-axis for each r-θ-cell, which serves as a distinctive velocity parameter for rotating target components. A subsequent determination of the azimuthal parameter for each (r,θ)-cell yields wheel positions and corresponding bulk velocities.

A micro-Doppler parameter $P_{\mu D}(r_i,\theta_j)$, quantifying the velocity deviation of all detections from mean velocity inside a 2-dimensional sliding window applied to each (r,θ)-cell, is defined as

$$P_{\mu D}(r_i,\theta_j) = \left(\frac{1}{A_{win}} \sum_{k,o,m} \left(v_m(r_{i+k},\theta_{j+o}) - \bar{v}(r_i,\theta_i)\right)^2\right)^{1/2}, \qquad (3.17)$$

with $k = -n_k/2, ..., +n_k/2$, $o = -n_o/2, ..., +n_o/2$ and $m = 1, ..., N_d(r,\theta)$, where n_k and n_o are the sliding window lengths in range and azimuth dimension, $N_d(r,\theta)$ is

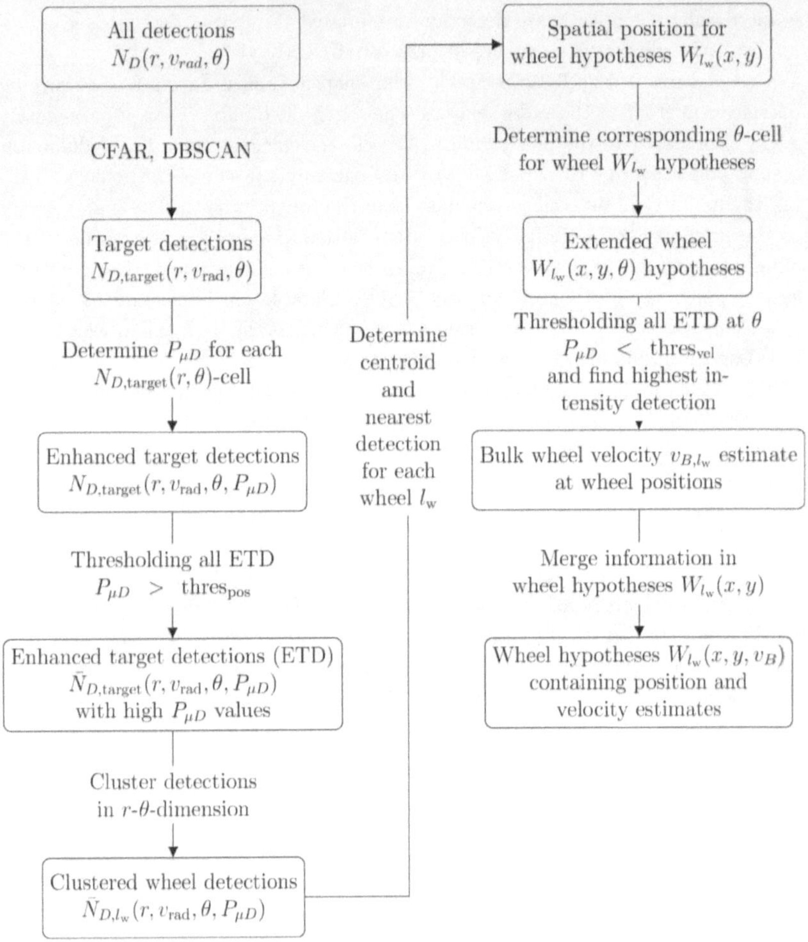

Figure 3.4: Flowchart to spatially resolve micro-Doppler spectra based on radar detections. The input are all target assigned detections of one radar measurement cycle. Output are wheel hypotheses $W_{l_{\text{w}}}(x, y, v_B)$ containing position and velocity estimates.

the number of reflection points inside the (k, o)-th cell, A_{win} is the area of the sliding window and $\bar{v}(r_i, \theta_i)$ is the mean velocity inside the sliding window. Eq. 3.17 determines the velocity variance within a sliding window for each (r, θ)-cell. The induced micro-Doppler signals presented in Fig. 3.3 indicate higher variance value, which predominantly occurs at wheel ranges in comparison to non-wheel ranges. Each target detection $N_{D,\text{target}}(r, v_{\text{rad}}, \theta)$ is extended by this micro-Doppler parameter, referred to as **enhanced target detections (ETD)**. Hence, a high $P_{\mu D}$-value is associated with elevated probability for a wheel located nearby. Deploying Eq. 3.17

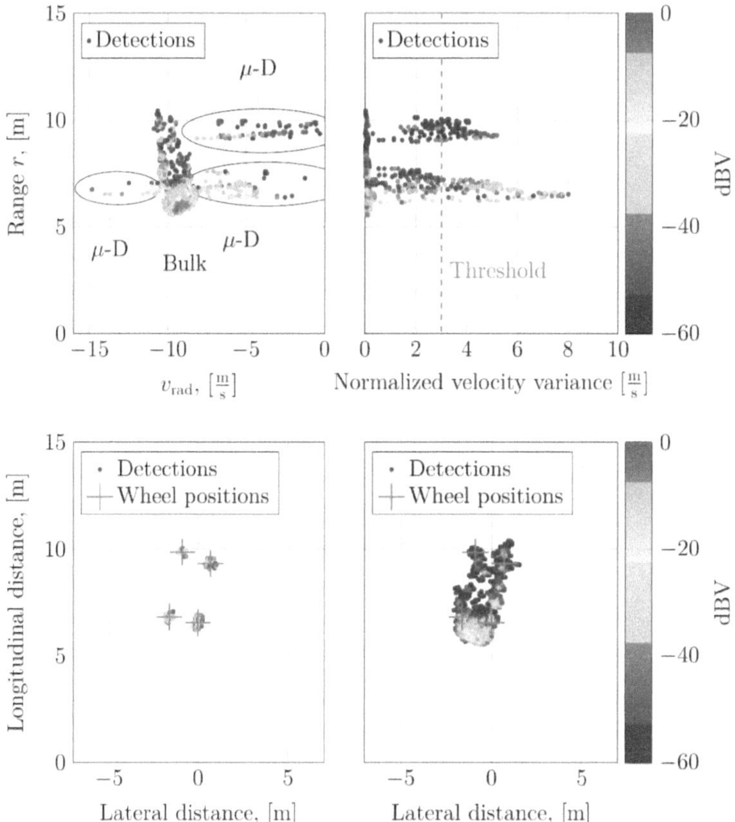

Figure 3.5: Radar data snapshot presenting the proposed method to determine the spatial location of the target vehicle during an evading scenario.
Top left: Snapshot of a target vehicle velocity profile, showing r-v_{rad}-detections, during an evading maneuver. The Doppler and micro-Doppler signals are clearly detectable. **Top right**: Visualization of the normalized velocity variance according to Eq. 3.17. Wheel induced r-v_{rad}-detections show large values. **Bottom left**: Result after the clustering procedure of high $P_{\mu D}$-valued detections. All threshold exceeding detections are considered in the clustering. The red crosses indicate the wheel positions and result from Eq. 3.18. **Bottom right**: Visualization of all detections in the x-y-plane with corresponding intensities. The red crosses indicate the wheel positions.

to all (r, θ)-detections and applying an additional threshold $P_{\mu D} > \mathrm{thres}_{\mathrm{pos}}$, the result separates bulk detections from wheel detections $\bar{N}_{D,\mathrm{target}}(r, v_{\mathrm{rad}}, \theta, P_{\mu D})$.
At this point, every (r, θ)-detection is associated with its new variance velocity parameter, the novel generated $P_{\mu D}$-value. A successive DBSCAN clustering to all points exceeding the threshold are grouped in the r-θ-dimension. These points are

referred to as **clustered wheel detections** and $\bar{N}_{D,l_{\mathrm{w}}}(r, v_{\mathrm{rad}}, \theta, \mu D)$ in Fig. 3.4, respectively. For each cluster, a centroid is determined using

$$W_{l_{\mathrm{w}}} = \frac{1}{N_{d,P_{\mu D}}} \sum_{N_{d,P_{\mu D}}} A_{d,P_{\mu D}}(x, y), \tag{3.18}$$

where $W_{l_{\mathrm{w}}}$ is the wheel position estimate, $N_{d,P_{\mu D}}$ is the number of clustered detections and $A_{d,P_{\mu D}}(x, y)$ are all clustered wheel detections $\bar{N}_{D,l_{\mathrm{w}}}(r, v_{\mathrm{rad}}, \theta, \mu D)$. These spatial information for all clustered wheel detections are stored in the wheel hypotheses $W_{l_{\mathrm{w}}}(x, y)$.

Fig. 3.5 visualizes the signal processing of the proposed method to determine wheel positions of a target vehicle executing an evading maneuver. The top left figure of Fig. 3.5 shows a snapshot of the target vehicle velocity profile, respectively r-v_{rad}-detections. The top right figure of Fig. 3.5 presents the normalized velocity variance. Wheel induced detection are likely to possess a broad velocity spread along the v_{rad}-axis. The potentially bulk generated detections are likely to range close to $0\,\frac{\mathrm{m}}{\mathrm{s}}$ whereas wheel induced detections are up to $8\,\frac{\mathrm{m}}{\mathrm{s}}$. The threshold value $\mathrm{thres_{pos}}$ intents to separate bulk-induced detections from wheel-induced detections. Since the bulk detections spread in the velocity dimension is range-dependent, as shown in the second and third row of Fig. 3.3, the threshold value is adjusted when the target vehicle is close to the sensor and generate broad bulk velocity spread.
The bottom left figure of Fig. 3.5 presents wheel detections and wheel position hypotheses result after clustering all detections possessing a high micro-Doppler parameter $P_{\mu D}(r_i, \theta_j)$. All threshold exceeding detections from the previous step are considered in the clustering step. The red crosses indicate the wheel positions. The final procedure results are visualized in the bottom right figure of Fig. 3.5 and show all detections in the x-y-plane with corresponding intensities. The red crosses indicate the wheel position hypotheses $W_{l_{\mathrm{w}}}(x, y)$.

To complete the parameter set, the Doppler velocity of the target at wheel locations are of interest. The velocity for the estimation of the target dynamics is the bulk motion of the vehicle at wheel positions, referred to as bulk wheel velocity. The expected Doppler velocity is independent of the range r [KSD16]. Hence, the Doppler velocity for a certain angle can be extracted from all scatterers from the considered direction.
Since there are more intense detections reflected from the vehicle body than from the wheels, an additional segmentation procedure is applied. First, all enhanced target detections (ETD) from the previous step with same angle θ as the respective wheel hypotheses $W_{l_{\mathrm{w}}}(x, y, \theta)$ are subject to $P_{\mu D} < \mathrm{thres_{vel}}$ identifying all detections which are predominantly originated from the vehicle body, since $P_{\mu D}$ is low. Second, for the remaining detections, the maximum value for measured intensity is taken and the corresponding **bulk wheel velocity** $v_{B,l_{\mathrm{w}}}$ of this single detection is determined. Fig. 3.6 shows the bulk wheel velocity $v_{B,l_{\mathrm{w}}}$ estimation in the θ-v_{rad}-plane. The

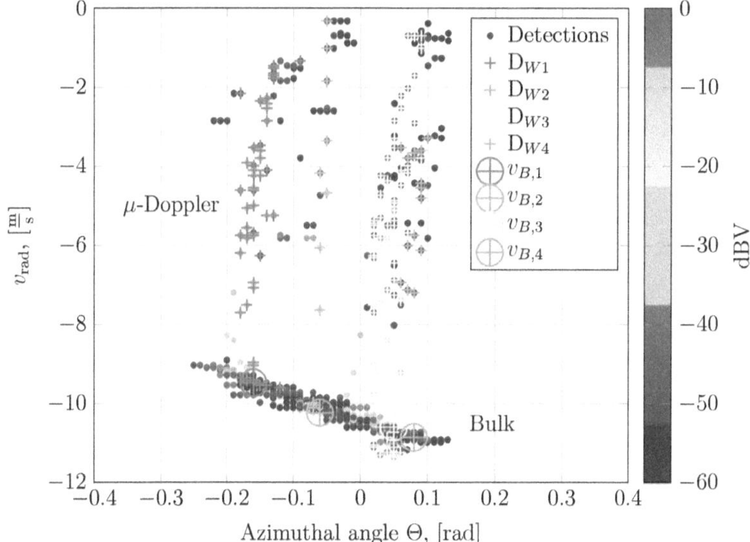

Figure 3.6: Snapshot presenting the procedure for velocity estimation per bulk wheel velocity (v_{B,l_w}), detections assigned to wheels based on the micro-Doppler parameter $P_{\mu D}(r,\theta)$ (D_{W1-4}) and every detection from the target vehicle during an evading maneuver in the Θ-v_{rad}-plane. The bulk velocities v_{B,l_w} are set to the most intense bulk detection for each angle segment containing a detected wheel. The bulk and micro-Doppler detections are clearly visible since the bulk detections accumulate at $-10\,\frac{\text{m}}{\text{s}}$.

snapshot presents the v_{B,l_w} velocity parameter, the detections assigned to wheels based on the micro-Doppler parameter $P_{\mu D}(r,\theta)$ (D_{W1-4}) and all detections from the target vehicle. Each cross indicates the Doppler and micro-Doppler detections assigned to one wheel, respectively. The most intense detection of all bulk detections per angular segment containing a wheel hypothesis is set as the bulk velocity value. The bulk and micro-Doppler detections are clearly visible since the bulk detections accumulate at $-10\,\frac{\text{m}}{\text{s}}$.

3.2.2 Validation using real radar data

The previous section introduces a generic target wheel positions and corresponding bulk velocity estimation procedure incorporating all target scattered detections and presents the method applied on a single measurement snapshot. This section evaluates the proposed method during a full maneuver with varying driving dynamics. Radar sensor measurements were conducted using a real vehicle as target object. The target is equipped with a real-time kinematic (RTK) reference measurement system to evaluate the wheel detection results. A GeneSys ADMA-G equipped with RTK positioning is used to obtain the ground truth global position

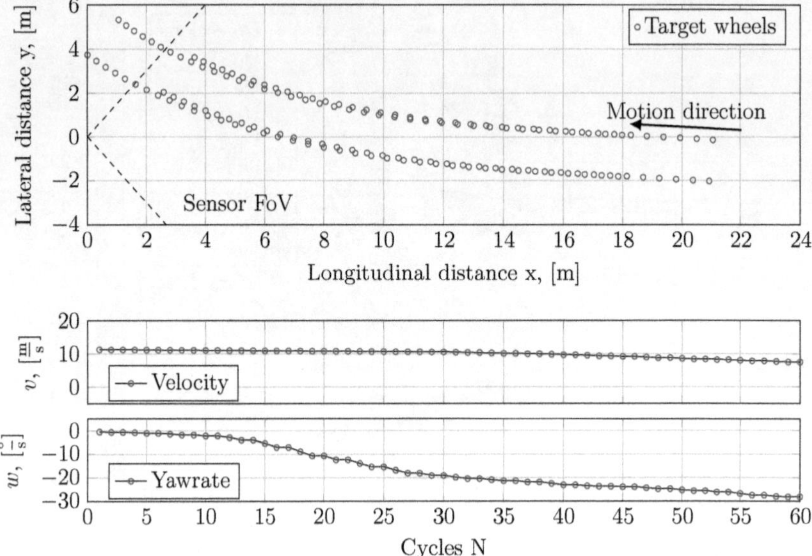

Figure 3.7: Target trajectory and dynamics during evading maneuver as scenario description.

estimate with an accuracy of $\pm 0.02\,\mathrm{m}$ and acceleration and turn rate data in three-axis $(a_x, a_y, a_z, w_x, w_y, w_z)$ for arbitrary positions of the target vehicle.

A high dynamic evading driving maneuver in front of the radar sensor is chosen as the evaluation scenario. The executed driving scenario is selected to evaluate the performance in low-range sensor-to-target configurations. Extensive range of yaw angles, yaw rates and safety-critical high longitudinal and lateral accelerations occur when the test driver tries to avoid an imminent collision in the very last moment.

Fig. 3.7 presents the proposed evading maneuver and its driving dynamics. The top figure of Fig. 3.7 visualizes wheel positions of the target vehicle seen from the birds-eye view and the bottom figure of Fig. 3.7 shows the target velocity and yaw rate for 60 measurement cycles. The radar sensor is located in the coordinate system's origin and the dashed lines indicate its FoV. The first measurement is taken at $12\,\mathrm{m}$, the initial target velocity is $10\,\frac{\mathrm{m}}{\mathrm{s}}$ and the target is heading straight at an initial yaw rate of $0\,\frac{\circ}{\mathrm{s}}$ without driver intervention. The vehicle is on a collision trajectory with the sensor. To avoid a collision, the driver initiates an evading maneuver without activating the braking system, causing an ascend of the yaw rate whereas the velocity is decreasing. In the further course of the maneuver, the yaw rate is increasing due to the high steering angle set by the driver and the velocity is decreasing. The target passes the sensor at a range of $2.2\,\mathrm{m}$ and avoids a collision

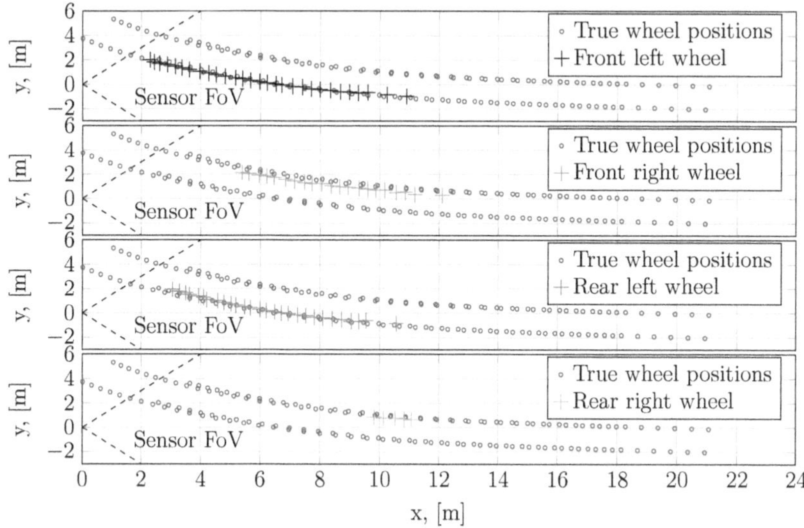

Figure 3.8: Spatially resolved micro-Doppler spectra based spinning wheel detection evaluation of wheel position during an evading maneuver.

with the sensor.

Fig. 3.8 presents the detected wheels for each respective wheel based on the spatially resolved micro-Doppler spectra evaluation. Fig. 3.9 shows the number N of detected wheels and bulk velocities at wheel positions compared to the true target velocities. Fig. 3.10 presents an error evaluation where estimated wheel positions and velocities are compared to reference data.

The first wheel (front-left, black marker) is detected at a range of approximately 12 m, which corresponds to cycle 17. Blue circles represent the true wheel positions and black crosses represent the estimated wheel positions, respectively. This wheel is detected almost during the entire maneuver. The accuracy for position, error ranging below 0.4 m, and velocity, error ranging below $2\,\frac{m}{s}$, is notably high since it is one of the direct LoS sensor-facing spinning wheel and hence likely to be detectable.

Two cycles later, the front right wheel is detected indicated by gray markers, respectively. The accuracy for position and velocity estimation is similarly compared to the front right wheel. However, the detection time is shorter than the left wheel due to a potential weaker micro-Doppler signal caused by an occlusion of the right wheel.

The rear left wheel is the third detected wheel, which corresponds to the red markers, respectively. The accuracy for position estimation is again similar up to cycle 45, followed by an increase up to 0.6 m for the lateral position. Notably is a decreasing accuracy of velocity estimation at the end of the maneuver.

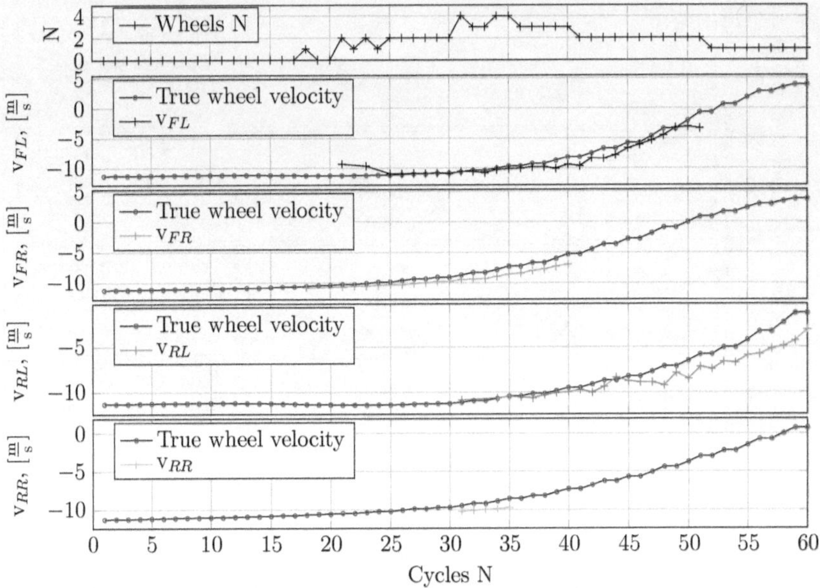

Figure 3.9: Spatially resolved micro-Doppler spectra based spinning wheel detection evaluation of bulk wheel velocities during an evading maneuver. The figures present the number N of detected wheels and the velocity estimation compared to the true target velocity for each detected wheel.

The rear right wheel corresponds to the green markers, respectively. The detection rate is comparably low due to extensive occlusion by the other wheels. This may be due to multipath propagation backscattered from the wheel and reflected between the vehicle floor and the road. However, the estimation accuracy for position and velocity are similar to the other wheels.

Fig. 3.10 presents the number of detected wheels per cycle, the errors for all estimated parameters compared to the reference sensor and the mean error for all parameters. The wheel number per cycle, once the full target extent is within $12\,\mathrm{m}$, ranges between $2-4$ wheels per cycle depending on the sensor-to-target configuration.

Overall the longitudinal position error e_x is comparably low, ranging below $0.4\,\mathrm{m}$, during the approaching and low yaw rate passage of the maneuver. It is slightly increased when the vehicle is reaching the sensor FoV boundaries. The lateral error e_y ranges up to approximately $0.8\,\mathrm{m}$ when the target is at the sensor FoV limits. The velocities errors are below $2\,\frac{\mathrm{m}}{\mathrm{s}}$, whereas the error increase with an increasing azimuthal angle under which the sensor illuminates the target. The additional target information is a promising parameter for a subsequent tracking procedure to estimate additional dynamic target states.

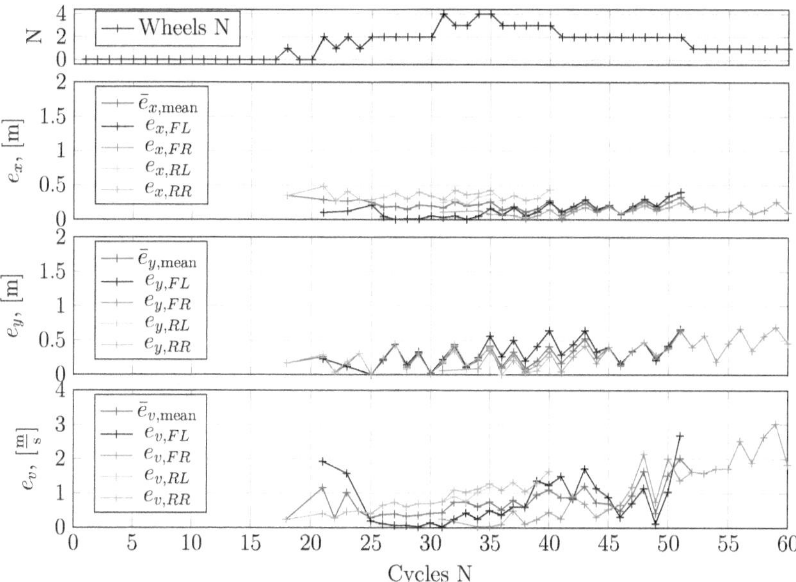

Figure 3.10: Error evaluation of spatially resolved micro-Doppler spectra based spinning wheel detection of wheel position and velocity during an evading maneuver compared to the reference sensor. The top figure shows the number N of detected wheels per cycle and the remaining figures present the mean longitudinal and lateral wheel position error and the velocity compared to the true target velocity at the wheel positions. The error $\bar{e}_x$, $\bar{e}_y$ and $\bar{e}_v$ show the mean value for all detected wheels for each cycle.

3.2.3 Conclusion

This chapter presents a novel procedure to estimate target wheel positions and corresponding bulk velocities based on radar sensor data. The contribution is to spatially resolve micro-Doppler signals, generated by rotating wheels of the target vehicle. The detected wheels serve as fixed scatter points on the target vehicle. Subsequently, characteristic points of the target are determined and the effect of wandering dominant reflection centers on the surface is mitigated. A micro-Doppler parameter is defined, which quantifies the velocity deviation of all detections from the mean velocity within a 2-dimensional sliding window applied to each (r, θ)-cell of the radar data cube. Since the velocity of each reflection at the target object is independent of its range, the velocity is set as the bulk velocity of the considered angle segment in which the respective wheel is present. The method is evaluated in a dynamic driving scenario where the driver performs an emergency evading action to avoid a collision with the sensor platform. The results proof feasibility of the proposed method and enable an advanced extended object tracking, incorporating

48

the additional target information, in the following chapter. Procedure improvement possibilities may be realized in the future to enhance overall detection performance by introducing sophisticated estimation methods for data segmentation and pre-processing, e.g., machine learning or curve fitting. Also a automated labeling of each wheel hypothesis is beneficial for the subsequent tracking procedure in future development stages.

3.3 Multipath propagation in uncertain environments

For convenience, a brief review of the multipath propagation problem in low-range sensor-to-target configurations is given and the section is based on [KHP+18] ©2018 IEEE. The challenge is the type of propagation that the radar signal may take on the transmitting or receiving path. After reflection on an obstacle, the coherently emitted waves of FMCW radar may be subject to multipath propagation [WHC16, GSYF17]. Then, the radar receives not only the direct reflection of an obstacle but also indirect temporally shifted reflection components. This superposition can lead to range-dependent interference patterns, which cause oscillating signal amplitudes of the received power [DKS+11]. In addition to performance degradation of the DoA estimation due to fading effects, multipath propagation can lead to the appearance of mirrored ghost targets, referred to as false-positive or ghost targets [VHZ18, EHZ+17].

Engels *et al.* [EHZ+17] showed that the occurrence of ghost targets through multipath reflections in long-range applications can lead to an overall deterioration of the sensor performance and creates high demands on signal processing algorithms concerning the suppression of mispredictions. By using a decoupled high-resolution frequency estimation method in the Fourier domain, estimation of the actual target positions that are subject to multipath propagation can be significantly increased. However, a detailed analysis of multipath propagation for correct classifications of ghost targets is still of primary importance. Taking into account the advanced driver assistance systems (ADAS) developments regarding large-scale integration of high-resolution, low-range radar sensors for direct vehicle environment, the influence of multipath reflections may increase significantly.

Hence, sensor environment perception contains non-existing obstacles, which can lead to fatal real-world driving situations for humans and vehicles in future auto-

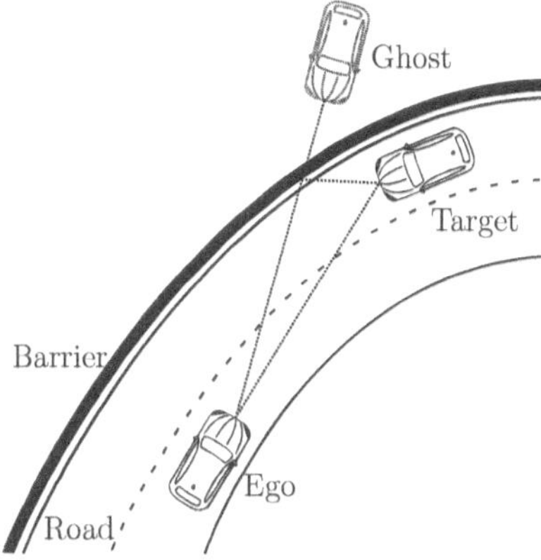

Figure 3.11: Driving scenario with ghost object presence [KHP+18] ©2018 IEEE.

mated driving scenarios.

Fig. 3.11 illustrates the effect on a typical traffic scenario. The ego vehicle is being approached by another vehicle along a curved road. The receiving radar signal consists of the corresponding propagation paths, including the specular multipath propagation via the guardrail inducing a ghost target with a velocity component directed towards the ego vehicle. This section presents a model-based geometric method that demonstrates the occurrence and the behavior of multipath reflections and validates the results with measured data of a custom and real vehicle target.

3.3.1 Multipath propagation model

Fig. 3.12 shows the multipath reflections that may occur on a reflection surface between a radar sensor and a target object at sensor height. The emitted electromagnetic wave is reflected by an object and is subject to multipath propagation. Consequently, the receiving radar signal represents a superposition of four propagation paths, whose frequency components assign different distances. This leads to the occurrence of false-positive detections $(O_{\mathrm{FP}\,n})$, where n is the number of false-positive detections. Note that the number of propagation paths is not limited to four. The considered paths and thereof resulting reflections are the predominant reflections.

This section presents a geometric multipath propagation model to determine the predominant electromagnetic wave paths and false-positive object positions. The sensor S travels on a circular path with radius $\vec{r}_m$ relative to a center point M and spans an angle ψ between the sensor S and the target position T. The electromagnetic wave gets reflected in reflection point P_{ref} at a distance $|\vec{r}_{\mathrm{ref}}|$ relative from center point M and angle $\psi/2$. Note that, additional information about the specular reflection surface are usually not available and thus, the assumption of the reflection point P_{ref} is at $\psi/2$ is made.

The distance between the target position T and the center point M is $|\vec{r}_t|$. For given M, target object radius $|\vec{r}_t|$, reflection point radius $|\vec{r}_{\mathrm{ref}}|$ and angle ψ, the target position T and the reflection point position P_{ref} can be determined using

$$\vec{r}_t = -\hat{r}_m \cdot \boldsymbol{X}_{\mathrm{rot}}^{\psi} \cdot |\vec{r}_t| + \vec{r}_m, \tag{3.19}$$

$$\vec{r}_{\mathrm{ref}} = -\hat{r}_m \cdot \boldsymbol{X}_{\mathrm{rot}}^{\frac{\psi}{2}} \cdot |\vec{r}_{\mathrm{ref}}| + \vec{r}_m, \tag{3.20}$$

where vectors are rotated by a rotation matrix $\boldsymbol{X}_{\mathrm{rot}}^{\alpha}$ by a clockwise angle α and $\hat{r}_m$ is the normalized vector.

One target T in the radar field of view generates three false-positive objects $O_{\mathrm{FP}\,1\text{-}3}$ due to multipath reflections. The target reflection is the direct LoS reflection. False-positive object $O_{\mathrm{FP}\,1}$ path origins from the sensor to the reflection surface, gets reflected towards the target object and back to the sensor. False-positive object $O_{\mathrm{FP}\,2}$ is the identical propagation path backwards. False-positive object $O_{\mathrm{FP}\,3}$ path starts at the sensor, gets reflected towards the target object, back to the reflection surface and back to the sensor leading to the faraway one.

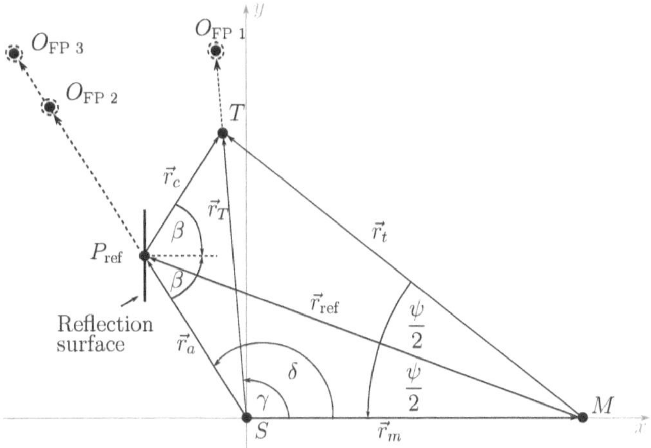

Figure 3.12: Multipath propagation model showing a radar sensor (S), a reflection point (P_{ref}) on a reflection surface, a target obstacle (T) and three false-positive reflections ($O_{\text{FP 1-3}}$).

The azimuthal angles spanned by $\angle(\vec{r}_m, \vec{r}_a)$ and $\angle(\vec{r}_m, \vec{r}_T)$, referred to as δ and γ, can be computed using vector analysis. The wave travel paths $|\vec{r'}_T|$ and $|\vec{r}_{\text{FP 1-3}}|$ can be determined using Eq. 3.19, Eq. 3.20 and

$$|\vec{r'}_T| = 2 \cdot (\underbrace{|\vec{r}_t - \vec{r}_m|}_{|\vec{r}_T|}), \tag{3.21}$$

$$|\vec{r}_{\text{FP 1,2}}| = \underbrace{|\vec{r}_{\text{ref}} - \vec{r}_m|}_{|\vec{r}_a|} + |\vec{r}_T| + \underbrace{|\vec{r}_t - \vec{r}_{\text{ref}}|}_{|\vec{r}_c|}, \tag{3.22}$$

$$|\vec{r}_{\text{FP 3}}| = 2 \cdot (|\vec{r}_a| + |\vec{r}_c|), \tag{3.23}$$

where $|\vec{r'}_T|$ is the propagation path length from the sensor to the target object and back, whereas $|\vec{r}_{\text{FP 1-3}}|$ are the path lengths from the sensor to the false-positive objects and back, respectively.

The position can be determined considering relative azimuth angles δ and γ when the electromagnetic waves get reflected back to the sensor either by the reflection point or by the target object with

$$\vec{O}_{\text{FP 1}} = \vec{e}_x \cdot |\vec{r}_{\text{FP 1}}| \cdot \boldsymbol{X}_{\text{rot}}^{-\gamma}, \tag{3.24}$$

$$\vec{O}_{\text{FP 2}} = \vec{e}_x \cdot |\vec{r}_{\text{FP 2}}| \cdot \boldsymbol{X}_{\text{rot}}^{-\delta}, \tag{3.25}$$

$$\vec{O}_{\text{FP 3}} = \vec{e}_x \cdot |\vec{r}_{\text{FP 3}}| \cdot \boldsymbol{X}_{\text{rot}}^{-\delta}, \tag{3.26}$$

where $\vec{e}_x$ is the unit vector in x-direction.

3.3.2 Multipath model validation using real radar data

Radar sensor measurements with a custom target, a real vehicle and a barrier are conducted to test and validate the proposed model. Fig. 3.13 presents the experimental setup. The radar sensor is placed stationary with an angle $\alpha = 45°$ rotated towards the reflection surface representing a vehicle corner radar. The custom target consists of two trihedral retroreflectors and a metal plate. One retroreflector causes a direct reflection back to the sensor. The other retroreflector reflects the wave, coming from the reflection surface, back to the barrier generating $O_{FP\,3}$. The metal plate mimics a license plate and the front bumper of a real vehicle and generates false-positive objects $O_{FP\,1}$ and $O_{FP\,2}$.

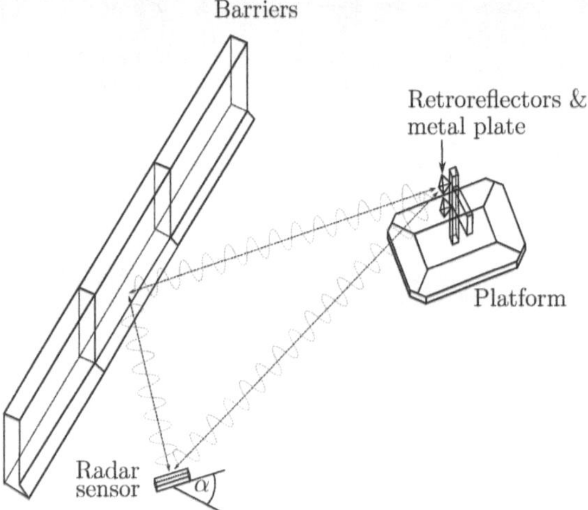

Figure 3.13: Experimental configuration showing the radar sensor, a barrier and the target object on a platform [KHP$^+$18] ©2017 IEEE.

The multipath propagation assumptions made in the model are validated using anechoic panels. The pyramidal shape and the material of the graphite coated polyurethane panels provide a significant attenuation and enable reproducible indoor radar testing. Measurements using the radar, custom target, real barriers and either with or without anechoic panels are carried out.

Furthermore, an intersection crossing scenario is constructed as a real-world urban driving scenario using the radar sensor as a corner radar. The sensor faces towards a building with an angle of 45° at a reasonable distance, e.g., sidewalk width. Another vehicle is approaching from the left and crossing the intersection straight. Fig. 3.14 and Fig. 3.15 show radar measurement results according to the described experimental setup, respectively.

Each figure shows the x-y-plane of the radar FoV with either the custom target or a real vehicle as present target and either a real barrier or a wall as reflection surface. In the following, the relative position of obstacles in the sensor FoV are referred to in Cartesian coordinates (x, y).

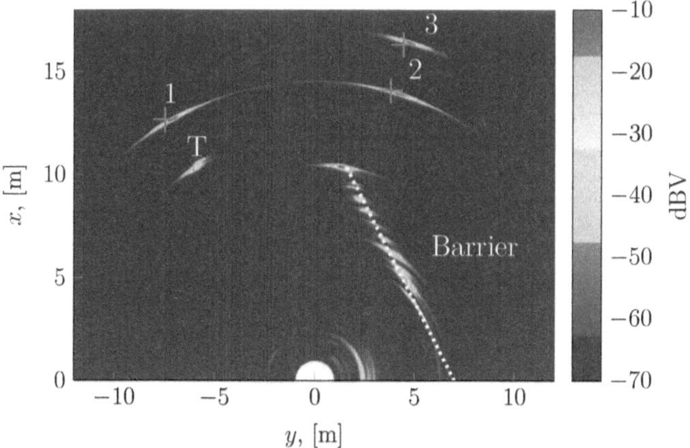

(a) Radar snapshot with a target (T) and 3 multipath reflections (1-3).

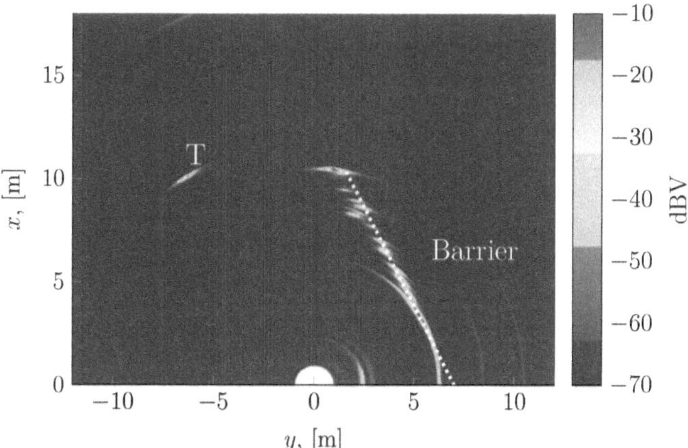

(b) Radar snapshot with mounted anechoic panels, the target and attenuated multipath reflections.

Figure 3.14: Multipath propagation model validation presenting radar snapshots with present target and a reflection surface. Red crosses indicate the ghost target positions determined using the model. The top figure shows results for blank barriers and the bottom figure shows results with covered barriers [KHP+18] ©2018 IEEE.

The top plot in Fig. 3.14 shows a single measurement snapshot illuminating blank barriers and the target object (T). Multipath reflections, indicated by 1, 2, and 3, from the barrier and the obstacle are clearly visible and appear as stated in section 3.3.1. The results of the presented model are marked as three red crosses localizing the corresponding false-positive multipath reflections. The bottom plot in Fig. 3.14 shows a measurement shot with the identical experimental setup illuminating barriers, but anechoic panels cover the direct reflection area. The target direct and

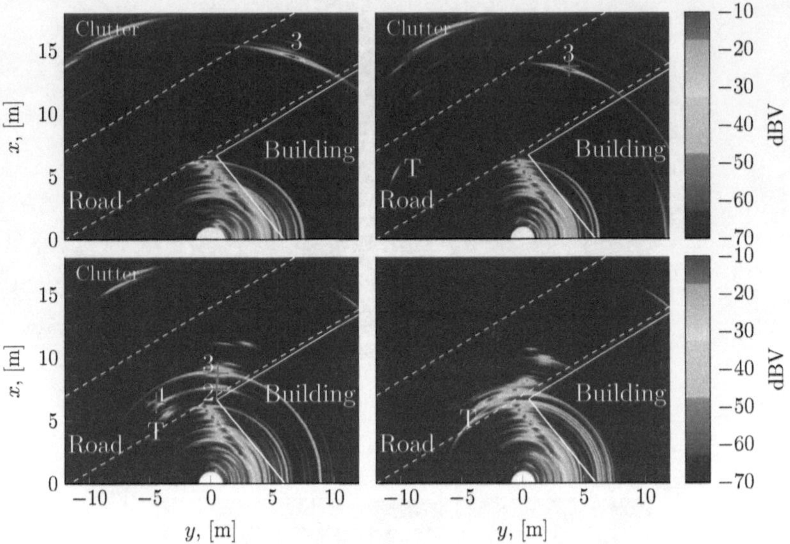

Figure 3.15: Radar shot results using a radar sensor, a custom reference target and a real target vehicle [KHP$^+$18] ©2018 IEEE.

barrier reflections are clearly visible, but the installation of several absorption panels attenuates all multipath reflections.

The top left plot in Fig. 3.15 shows a building as reflection surface in close distance to the sensor causing intense reflections and broad peaks. In the upright corner, at (6, 14.5), one multipath reflection from the real vehicle as the target is present. It corresponds to $O_{FP\ 3}$ from the model. The rotated radar sensor does not receive a direct reflection from the target vehicle. The target is to the left of the image, and hence, undetectable in this early stage of the scenario.

The top right plot in Fig. 3.15 show both, the target vehicle and one multipath reflection in the same scenario a few cycles later. False-positive reflections are more intense due to the radar sensor antenna characteristic emitting a high power level in the main lobe and less power in the side lobes. The multipath reflection can be identified as $O_{FP\ 3}$ from the model.

The bottom left plot in Fig. 3.15 shows the target vehicle, which has further approached to the sensor. The false-positive reflections and the LoS reflection from the vehicle decomposed into multiple intense peaks. Again, the false-positive reflection shows a higher intensity due to the radar sensor antenna characteristic. The radar sensor illuminates the target vehicle right side leading to the less intense reflection (T) and simultaneously enables multipath reflections $O_{FP\ 1\text{-}3}$. Due to the close distance from the sensor, reflection surface, target vehicle and, in addition, the large reflection areas from the building and the target vehicle side surface, it is highly probable to generate all multipath reflections at the same time.

The bottom right plot in Fig. 3.15 shows the target vehicle which is at the same lateral position as the wall of the building causing a random generation of multipath and direct LoS reflections from the radar sensor to reflection surface and backward to the sensor. A separation between multipath reflections and the true target vehicle is hardly possible.

3.3.3 Conclusion

The presented model analyzes the emergence and behavior of radar-based multipath reflections, leading to multiple false-positive targets and an elevated false alarm rate in real-world driving scenarios. Based on propagation and reflection behavior of electromagnetic waves, a geometric reflection model is derived illustrating the occurring multipath reflections in real-world surfaces, e.g., buildings or road-bounding barriers, e.g., guardrails. The geometric propagation model determines the relative positions of the false-positive reflections and validated them with extensive radar sensor data. A custom reflector target mounted on a platform, creating deterministic point targets as dominant backscatter centers of a vehicle body, validated the different multipath reflections and the overall accuracy of the model. Radar measurements of a vehicle during an intersection driving scenario provides detailed insights into the multipath reflection behavior and additionally confirms the assumptions made in the model. The model may be enhanced in employing a reflection surface model. Thereby the simplification assumption that the reflection point is at $\psi/2$ may become redundant.

The results identify the emerging challenges for the correct interpretation of false-positive targets to enable sophisticated automotive safety-, comfort- and automation applications. Based on these results, further investigations on the effects of multipath reflections for automotive radar sensors need to be carried out to analyze the influence also on the object velocity determination based on the Doppler effect. Observations show a shifted velocity signal for every false-positive object. Note that the multipath model does not apply in the tracking chapter. This effort is left for future work.

3.4 Summary

This chapter presented the fundamentals of electromagnetic wave propagation for the Fresnel and Fraunhofer region, the micro-Doppler effect and introduced two novel methods – spinning wheel detection based on spatially resolved micro-Doppler signals and a generic multipath propagation model to identify false-positive reflections.

The micro-Doppler effect of vibrating or rotating target components, e.g., spinning wheels, is detectable in close sensor-to-target configurations and delivers additional target information. The micro-Doppler effect for spinning wheels was derived and real radar data snapshots presented the measurable micro-Doppler effect in approaching scenarios. These Doppler signals are used to determine wheel position and bulk velocity as characteristic points. The characteristic points are used to solve the problem of wandering dominant scatter points on the vehicle surface in the next chapter. The scientific contribution is to spatially resolve the micro-Doppler effect to determine the characteristic features. A micro-Doppler parameter is defined, which quantifies the velocity deviation of all detections from the mean velocity inside a sliding window applied to each range-angle-cell. This method is evaluated in a dynamic evading maneuver. The test driver performs an emergency evading action to avoid a collision with the sensor platform. The results proof feasibility of the proposed method and enable a subsequent extended object tracking, incorporating the additional target information, in the following chapter.

In comparison to the detections reflected from the object of interest, target vehicles in close distance may be subject to multipath propagated detections, where the transmitted electromagnetic wave potentially gets reflected by other objects or structure surfaces, e.g., guardrails. This chapter introduced a survey for the number and position of multipath detections and a generic method to identify false-positive objects. The presented model is based on computation efficient geometric vector analysis for given radar detections. Besides this model and to the best of the authors knowledge, the survey is the first scientific analysis that proved the generation of multiple ghost objects per present target. The relative positions of generated multipath objects depend on the specific circular path traveled by the electromagnetic wave. The utilization of a custom reflector and real vehicle as target validated the different multipath reflections and the overall accuracy of the model.

References

[CLHW03] Victor C. Chen, Fayin Li, Shen-Shyang Ho, and Harry Wechsler. Analysis of micro-Doppler signatures. *IEE Proceedings-Radar, Sonar and Navigation*, 150(4):271–276, 2003.

[CLHW06] Victor C. Chen, Fayin Li, Shen-Shyang Ho, and Harry Wechsler. Micro-Doppler effect in radar: Phenomenon, model, and simulation study. *IEEE Transactions on Aerospace and Electronic systems*, 42(1):2–21, 2006.

[DKH⁺15] Jürgen Dickmann, Jens Klappstein, Markus Hahn, Marc Muntzinger, Nils Appenrodt, Carsten Brenk, and Alfons Sailer. Present research activities and future requirements on automotive radar from a car manufacturer's point of view. In *2015 IEEE MTT-S International Conference on Microwaves for Intelligent Mobility (ICMIM)*, pages 1–4. IEEE, 2015.

[DKS⁺11] Fabian Diewald, Jens Klappstein, Frederik Sarholz, Jürgen Dickmann, and Klaus Dietmayer. Radar-interference-based bridge identification for collision avoidance systems. In *2011 IEEE Intelligent Vehicles Symposium (IV)*, pages 113–118. IEEE, 2011.

[EHZ⁺17] Florian Engels, Philipp Heidenreich, Abdelhak M. Zoubir, Friedrich K. Jondral, and Markus Wintermantel. Advances in automotive radar: A framework on computationally efficient high-resolution frequency estimation. *IEEE Signal Processing Magazine*, 34(2):36–46, 2017.

[GSYF17] Min Guo, Zhanshan Sun, Shilin Yang, and Yunqi Fu. Genetic algorithm improved spatial diversity 24-GHz FMCW radar with multipath for automotive applications. In *2017 Progress in Electromagnetics Research Symposium-Fall (PIERS-FALL)*, pages 1988–1993. IEEE, 2017.

[HLS12] Lars Hammarstrand, Malin Lundgren, and Lennart Svensson. Adaptive radar sensor model for tracking structured extended objects. *IEEE Transactions on Aerospace and Electronic Systems*, 48(3):1975–1995, 2012.

[HSSS12] Lars Hammarstrand, Lennart Svensson, Fredrik Sandblom, and Joakim Sorstedt. Extended object tracking using a radar resolution model. *IEEE Transactions on Aerospace and Electronic Systems*, 48(3):2371–2386, 2012.

[KBK⁺15] Dominik Kellner, Michael Barjenbruch, Jens Klappstein, Jürgen Dickmann, and Klaus Dietmayer. Wheel extraction based on micro Doppler distribution using high-resolution radar. In *2015 IEEE MTT-S International Conference on Microwaves for Intelligent Mobility (ICMIM)*, pages 1–4. IEEE, 2015.

[Kel17] Dominik Kellner. *Verfahren zur Bestimmung von Objekt-und Eigenbewegung auf Basis der Dopplerinformation hochauflösender Radarsensoren.* PhD book, Universität Ulm, 2017.

[KHP+18] Alexander Kamann, Patrick Held, Florian Perras, Patrick Zaumseil, Thomas Brandmeier, and Ulrich T. Schwarz. Automotive radar multipath propagation in uncertain environments. In *2018 21st International Conference on Intelligent Transportation Systems (ITSC)*, pages 859–864. IEEE, 2018.

[KSD16] Christina Knill, Alexander Scheel, and Klaus Dietmayer. A direct scattering model for tracking vehicles with high-resolution radars. In *2016 IEEE Intelligent Vehicles Symposium (IV)*, pages 298–303. IEEE, 2016.

[LDL11] Yanbing Li, Lan Du, and Hongwei Liu. Moving vehicle classification based on micro-Doppler signature. In *2011 IEEE International Conference on Signal Processing, Communications and Computing (ICSPCC)*, pages 1–4. IEEE, 2011.

[Mah05] Bassem R. Mahafza. *Radar Systems Analysis and Design Using MATLAB Second Edition.* Chapman and Hall/CRC, 2005.

[PLH92] Kevin J. Parker, Robert M. Lerner, and Sung-Rung Huang. Method and apparatus for using Doppler modulation parameters for estimation of vibration amplitude, February 11 1992. US Patent 5,086,775.

[VHZ18] Tristan Visentin, Jürgen Hasch, and Thomas Zwick. Analysis of multipath and DOA detection using a fully polarimetric automotive radar. *International Journal of Microwave and Wireless Technologies*, 10(5-6):570–577, 2018.

[WHC16] Hsiao-Ning Wang, Ying-Wei Huang, and Shyh-Jong Chung. Spatial diversity 24-GHz FMCW radar with ground effect compensation for automotive applications. *IEEE Transactions on Vehicular Technology*, 66(2):965–973, 2016.

[ZRH98] Mark S. Zediker, Robert R. Rice, and Jack H. Hollister. Method for extending range and sensitivity of a fiber optic micro-Doppler ladar system and apparatus therefor, December 8 1998. US Patent 5,847,817.

Chapter 4

Target tracking in pre-cash scenarios

The focus of this chapter is on principles of target tracking in pre-crash scenarios. Thus, the chapter provides a brief but comprehensive overview of Bayesian state estimation. It subsequently introduces three fundamental filter designs: the Kalman filter (KF) for linear process and measurement models, the extended Kalman filter (EKF) for nonlinear process and measurement models and the particle filter (PF), which is well-suited to represent complex multimodal beliefs and relax the requirement for strong parametric assumptions on the posterior densities [TBF05].

As motivation and problem formulation, the first section presents a real-world survey to evaluate the performance of a series radar sensor. The evaluation is conducted using a novel test method. The vehicle with the mounted radar sensor is transferred in a skid situation to reproducibly generate dynamic, pre-crash scenarios. In these situations, the driver attempts to avoid imminent collisions by inducing high steering wheel angles and braking forces. The results provide evidence of the need for research and development efforts for methods and implementations in this specific driving situations to ensure a seamless and accurate target state estimation in skid situations.

The contributions in this chapter include the novel test method with an extended nonlinear vehicle motion model and a subsequent EKF implementation. The motion model is used to estimate parameters of the horizontal vehicle motion during a skid event. An EKF is designed to track a static target object using pre-processed automotive radar signals as input. The novel filter is tested and validated under realistic conditions using a test vehicle equipped with state-of-the-art automotive sensors in skid situations. The filter estimates are compared to the true target motion measured with a reference system.

Another contribution in this chapter deals with an uncertain measurement model problem. The wandering dominant scatter point on the extended object surface leads to additional uncertainty, which needs to be considered. A subsequent tracking procedure, incorporating target wheel hypotheses from the previous chapter 3, is designed to provide a solution to the problem mentioned above. The positions and bulk velocities of spinning target wheels serve as fixed and characteristic points on the vehicle surface and are suitable for the subsequent tracking. The signal processing and tracking procedure estimates the position and corresponding bulk velocity for near objects evaluating micro-Doppler signals, generated by rotating target wheels. This novel information serves as input to a particle-filter based tracking framework. The presented approach covers signal processing from raw radar sensor data to extended target state estimation. The performance is evaluated on real experimental data where a target vehicle with a mounted reference sensor is used to reconstruct safety-critical, dynamic evading maneuvers.

4.1 Test method for environment sensors in skid scenarios

A brief survey of a test vehicle with mounted automotive radar sensor in skid situations is given to emphasize and motivate the following sections. Accordingly, a novel test method is developed to test automotive environment perception sensors in a reproducible and non-destructive manner. Reproducible tests decrease the test effort and support the test-driven development of new safety systems. This section is based on [KBH+17] and [KHD+17].

The test method utilizes a kick plate, which applies a lateral force to the rear axle of the ego-vehicle. An exemplary ego-vehicle trajectory is shown in Fig. 4.1. Kick plates are used to improve driver reactions in unexpected skid situations to avoid over- or understeering and the driver is trained to transfer the vehicle back in a safe state [ABBP+11].

The ego-vehicle approaches a detected target obstacle and overruns the kick plate, which is located between ego-vehicle and target obstacle. An external lateral force $\vec{F}_{kick}$ is applied on the rear axle by the kick plate. This force induces a rotational component to the translational motion. It affects the vehicle yaw rate w and side slip angle (SSA) β. As a consequence, the ego-vehicle yaw rate increases. The vehicle is transferred in a skid driving situation under proximate deterministic conditions. The yaw rate depends on the magnitude of external force $\vec{F}_{kick}$ and leads to increased SSA, changing the direction of the longitudinal vehicle axis relative to its motion direction. Sensor accuracy and limitations can be analyzed and measured in a controlled manner by incrementing the shifting intensity of the kick plate. The detection, classification and tracking performance for various surround sensors can be tested by variation of the target position and the magnitude of the external force. For the proposed survey, a test vehicle was used to record sensor data in regular and skid driving situations. The vehicle, with a mounted ARS430 series radar sensor and a RTK reference sensor, is used to carry out the sensor performance tests under realistic and reproducible test conditions. The sensors are connected to a measurement computer in the trunk and the sensor data are recorded. However, the sensor provides a sophisticated object list as well as pre-processed raw data. In close distance to the target object, a single measurement cycle may provide several detections, similar to the development presented radar. The scattered radar echoes

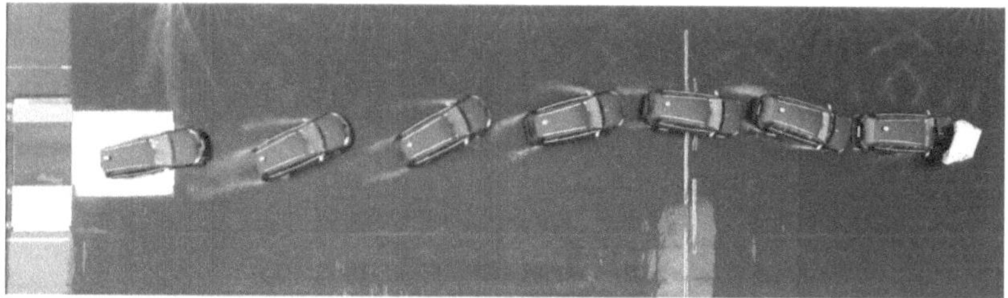

Figure 4.1: Sequence of a dynamic test run taken with a drone; time step: 0.5 s [KHD$^+$17] ©2017 IEEE.

are received and further processed to clustered detections with relative distance, bearing and radial velocity parameters. These grouped detections are the output of the sensor raw signal processing (RSP). For one object, the sensor may generate multiple grouped detections, referred to as clusters. The measurements were carried out with a standardized euro vehicle target (EVT) as a realistic radar target, which can be crashed without damaging the test vehicle. This target is used to test and validate various safety functions, e.g., emergency brake assist (EBA). Details of the radar target can be found in [San13]. For the proposed skid scenarios, a distinction between regular driving situations and skid driving situations is introduced. In regular driving situations, no external force is applied and the test vehicle approaches the EVT straight. Ego-vehicle velocities range between 30 km/h and 50 km/h without significant steering inputs. However, for skid scenarios, the kick plate is used to transfer the vehicle and its sensors in skid driving situations. The SSA is taken as the criticality criterion to distinguish between regular and skid scenarios. For regular test runs, the maximal SSA is smaller than 5° and for skid scenarios, the SSA is larger than 5°, respectively.

Each test run includes three stages: the vehicle is accelerated to an initial velocity, the rear axle is laterally shifted by the kick plate causing the vehicle to increase its yaw angle and SSA while sliding across the watered area with low friction. Eventually, the test driver attempts to stabilize the vehicle while it is skidding towards the target. Fig. 4.1 shows a top view sequence taken with a drone. Several test run videos can be found at **https://goo.gl/kREjEU**.

Fig. 4.2 shows various radar sensor parameter measuring the stationary EVT during a skid scenario with 57 cycles. For the reader convenience, the ego-vehicle dynamics measured by the reference sensor for yaw rate, yaw angle and SSA are shown in the top figure. The second and third-row figures present the radar sensor position and velocity estimation indicated with crosses and corresponding reference parameters indicated with black squares and circles, respectively. The $\pm 2\sigma$ standard deviation for each sensor signal is represented by the gray shades engulfing the graphs, respectively. The bottom figure presents two object qualifier: the obstacle probability and the existence probability. The first parameter is a estimate if an object is overrunable and the second parameter is a measure for the object existence probability. According to the top figure, the kick plate is activated after the third measurement

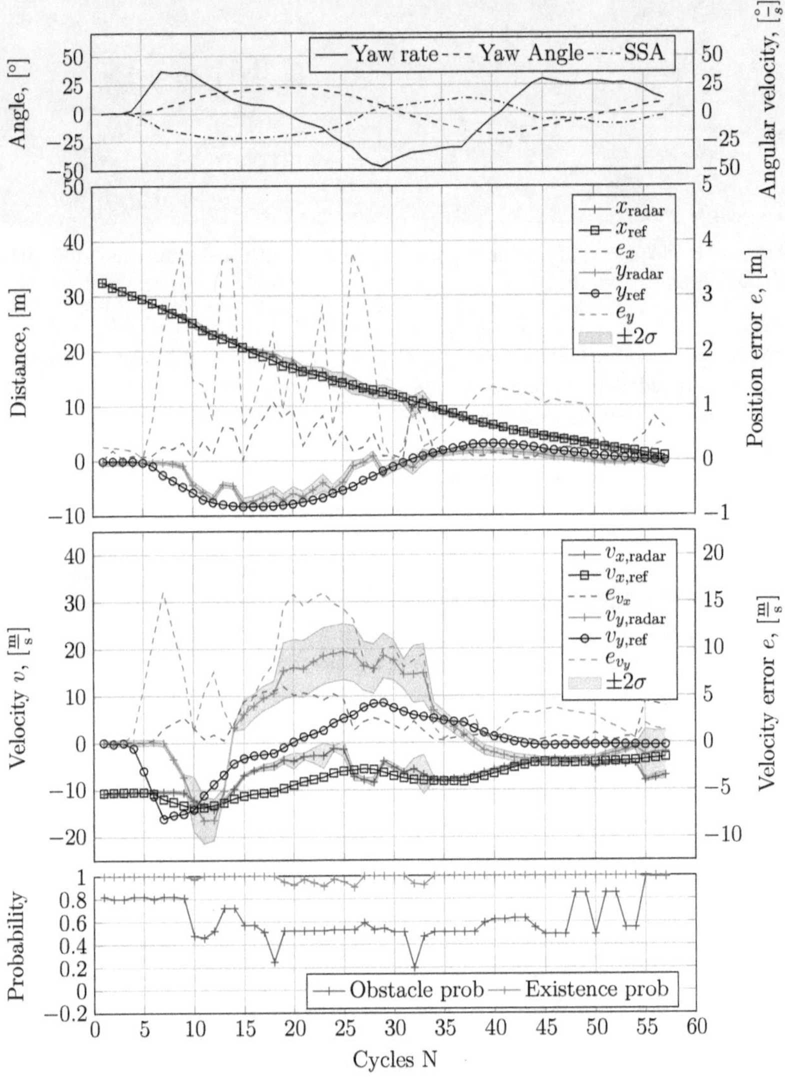

Figure 4.2: Radar sensor performance survey, presenting radar sensor parameter measuring a stationary EVT, during a skid test run with 57 cycles. The top figure shows the ego-vehicle dynamic parameters: yaw rate, yaw angle and SSA. The second and third row figures show the radar sensor position and velocity estimation indicated with crosses, the reference values, indicated with black squares and circles, and the $\pm 2\sigma$ standard deviation for each radar sensor signal represented by the gray shades, respectively. The bottom figure presents two object qualifier: the obstacle and existence probability. The first parameter is a estimate if an object is overrunable and the second parameter is a measure for the object existence probability.

cycle as the yaw rate, yaw angle and side slip angle start to increase. The yaw rate, indicated by a solid line, rises to $36\,\circ/s$ illustrating the high dynamic driving situation. The SSA, indicated by a dashed-dotted line, exceeding $-20\,°$ and proofs the skid driving situation. During the scenario, the yaw angle rises to $25\,°$. The test driver attempts to stabilize the vehicle and oscillates between under- and oversteering situations leading to alternating SSA and yaw angle. The relative x-y-position and -velocity show an increased deviation to the reference. The maximum absolute error is $1\,m$, whereas the lateral distance error ranges up to $3.8\,m$ during the skid situation. The errors for each signal are indicated as dashed blue and red curves, respectively.

The corresponding velocity estimation errors are presumably in similar or larger ranges. The errors range up to $6\,m/s$ and $16\,m/s$ for longitudinal and lateral velocity estimations. Remarkable are the large lateral estimation error values right after the kick plate event and during the first half of the skid scenario with a corresponding wide $\pm 2\sigma$-band. Again, the errors for each signal are indicated as dashed blue and red curves, respectively. The bottom figure shows a significant reduction of the existence and obstacle probabilities during the skid event, proofing a decrease in the environment perception accuracy and potential misinterpretations.

Concluding, the proposed survey proved temporarily large deviations up to $3.8\,m$ and $16\,m/s$ to the actual position and velocities during skid events, respectively. Hence, a tracking framework in skid situations is left to be developed.

4.2 Extended Kalman filter based point target state estimation in skid scenarios

4.2.1 Bayesian state estimation

The Bayesian state estimation, also referred to as object or target tracking, is based on the recursively estimated posterior probability of the target states incorporating all available observations of the target. In the automotive environment perception context, the information of interest is, e.g., number of present objects, static and dynamic object states. Generally speaking, object tracking aims to estimate the probability density function (PDF) $p(\mathbf{x})$ of states $\mathbf{x}$ by integrating the available object information over time. The probability density can be expressed as

$$p(\mathbf{x}_k) = p(\mathbf{x}_k|\mathbf{Z}_k), \tag{4.1}$$

where $k \in \mathbb{N}$ is the time index, $\mathbf{x}_k \in \mathbb{R}^{n_x}$ is the target state vector, n_x the dimension of the state vector and $\mathbf{Z}_k$ contains all received measurements up to that time. For this work, we restrict ourselves to signals modelled as Markovian, nonlinear, non-Gaussian state-space models. Hence, the target states evolve according to a known Markov model represented by f_{k-1}, e.g., a physical motion model of an object

$$\mathbf{x}_k = f_{k-1}(\mathbf{x}_{k-1}, \mathbf{v}_{k-1}). \tag{4.2}$$

Note that different objects may have different motion models according to their dynamics. Markov models may be linear or nonlinear motion models, e.g., the presented single track motion model in section 4.2.3. It is essential to notice that the future state may be stochastic, but no variables prior to $\mathbf{x}_{k-1}$ may influence the stochastic evolution for future states unless this dependence is mediated through the state $\mathbf{x}_{k-1}$. Markov chains meet these temporal processes conditions [TBF05]. The relation between the target state and the measurements is described by measurement function h_k

$$\mathbf{z}_k = h_k(\mathbf{x}_k, \mathbf{w}_k), \tag{4.3}$$

where $\mathbf{z}_k \in \mathbb{R}^{n_z}$ is the measurement vector with n_z as measurement vector dimension and $\mathbf{v}_{k-1}$ and $\mathbf{w}_{k-1}$ are white, independent process and measurement noises. The noises represent the model uncertainty due to model approximations and sensor-specific measurement uncertainties, respectively. The measurement function translates the states from state space to measurement space, e.g., when a position is translated from a vehicle coordinate system to a sensor coordinate system to enable a mathematical operation.

The posterior density function $p(\mathbf{x}_k|\mathbf{Z}_k)$ can be recursively obtained using the prediction and update stages [GRA04]. Assuming the posterior density $p(\mathbf{x}_{k-1}|\mathbf{Z}_{k-1})$ at time step $k - 1$ is available, where $\mathbf{Z}_{k-1}$ contains all received measurements up

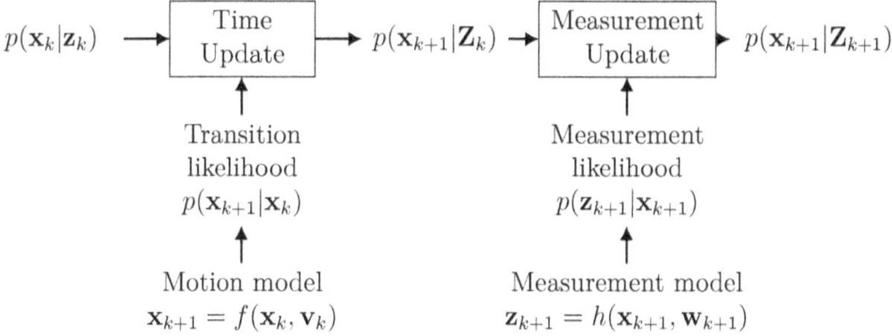

Figure 4.3: Bayesian state estimation principles based on [GLGO14].

to that time, the prior density (predicted density) at time k may be obtained from the system model via the Chapman-Kolmogorov equation

$$p(\mathbf{x}_k|\mathbf{Z}_{k-1}) = \int p(\mathbf{x}_k|\mathbf{x}_{k-1})p(\mathbf{x}_{k-1}|\mathbf{Z}_{k-1})d\mathbf{x}_{k-1}, \tag{4.4}$$

where use has been made of the fact that the vehicle motion model is a first-order Markov process and the term simplifies to $p(\mathbf{x}_k|\mathbf{x}_{k-1}, \mathbf{Z}_{k-1}) = p(\mathbf{x}_k|\mathbf{x}_{k-1})$. The next stage updates the prior density. It is carried out when a new measurement $\mathbf{z}_k$ becomes available using the Bayes' rule

$$p(\mathbf{x}_k|\mathbf{Z}_k) = p(\mathbf{x}_k|\mathbf{z}_k, \mathbf{Z}_{k-1}), \tag{4.5}$$

$$= \frac{p(\mathbf{z}_k|\mathbf{x}_k, \mathbf{Z}_{k-1})p(\mathbf{x}_k|\mathbf{Z}_{k-1})}{p(\mathbf{z}_k|\mathbf{Z}_{k-1})}, \tag{4.6}$$

$$= \frac{p(\mathbf{z}_k|\mathbf{x}_k)p(\mathbf{x}_k|\mathbf{Z}_{k-1})}{p(\mathbf{z}_k|\mathbf{Z}_{k-1})}, \tag{4.7}$$

where the denominator

$$p(\mathbf{z}_k|\mathbf{Z}_{k-1}) = \int p(\mathbf{z}_k|\mathbf{x}_k)p(\mathbf{x}_k|\mathbf{Z}_{k-1})d\mathbf{x}_k, \tag{4.8}$$

depends on the likelihood function $p(\mathbf{z}_k|\mathbf{x}_k)$, defined by the measurement model. The recurrence forms the optimal Bayesian solution and requires a prior density $p(\mathbf{x}_0)$ at $k = 0$.

Practical considerations: Fig. 4.3 shows an exemplary time and measurement update cycle and illustrates the coherence of motion and measurement models in Bayesian state estimation. The figure and exemplary description is based on [GLGO14]. A Bayesian state estimation algorithm composes two stages: the time update and the measurement update. In this example, the single state $\mathbf{x}_k$ generates a single detection $\mathbf{z}_k$ in each discrete time step k. The objective of single-object estimation is to estimate the object state $\mathbf{x}_k$ by using $\mathbf{z}_k$. The time update stage predicts the motion that the object performs between two observations. In most cases, the motion is unknown and, hence, simplistic premises can be made about the motion that the object is performing. To mathematically describe the motion, a motion model of the form $\mathbf{x}_{k+1} = f(\mathbf{x}_k, \mathbf{v}_k)$ may be used. Random process noise $\mathbf{v}_k$ handles uncertainties and imperfect modeling. The measurement update uses the observations delivered by the sensor to update the object estimate. Commonly, a measurement model of the form $\mathbf{z}_{k+1} = h(\mathbf{x}_{k+1}, \mathbf{w}_{k+1})$ is used.

Each observation is corrupted by noise $\mathbf{w}_k$. Each time step k, the sensor delivers one observation $\mathbf{z}_k$ and $\mathbf{Z}_k$ contains all observations from time 1 to time k. Due to the uncertainties involved, e.g., measurement and process noise, the knowledge of the object state $\mathbf{x}$ is often described using probability distributions $p(\mathbf{x}_k|\mathbf{Z}_k)$. The evolution of the distribution of state $\mathbf{x}$ over time can be described using appropriate measurement and motion models $h(\cdot)$ and $f(\cdot)$, respectively. In the following section an extended Kalman filter (EKF) and a particle filter (PF) framework will be used to determine the posterior densities. Note that both approaches are based on the presented Bayesian state estimation principles.

4.2.2 Extended Kalman filter

The extended Kalman filter (EKF) is based on the Kalman filter (KF), a technique for filtering and prediction of linear Gaussian systems invented by Swerling (1958) and Kalman (1960) [Kal60]. The following notations rely on [TBF05] and a detailed mathematical derivation of the KF can be found also in [TBF05].

One of the most popular state estimation techniques is the KF, which is applicable for linear Gaussian systems. Filter results are obtained by a set of recursive prediction and update equations. The results are represented by beliefs at time k with mean $\boldsymbol{\mu}_k$ and covariance $\mathbf{P}_k$. The KF relies on the following linearity assumptions for state transition probability $p(\mathbf{x}_k|\mathbf{x}_{k-1})$ and measurement probability $p(\mathbf{z}_k|\mathbf{x}_k)$, which can be expressed as multivariate normal distributions

$$p_{\mathrm{KF}}(\mathbf{x}_k|\mathbf{x}_{k-1}) = \det(2\pi\mathbf{Q}_k)^{-\frac{1}{2}} \exp\left(-\frac{1}{2}(\mathbf{x}_k - \mathbf{F}_k\mathbf{x}_{k-1})^T \mathbf{Q}_k^{-1}(\mathbf{x}_k - \mathbf{F}_k\mathbf{x}_{k-1})\right), \quad (4.9)$$

$$p_{\mathrm{KF}}(\mathbf{z}_k|\mathbf{x}_k) = \det(2\pi\mathbf{R}_k)^{-\frac{1}{2}} \exp\left(-\frac{1}{2}(\mathbf{z}_k - \mathbf{H}_k\mathbf{x}_k)^T \mathbf{R}_k^{-1}(\mathbf{z}_k - \mathbf{H}_k\mathbf{x}_k)\right). \quad (4.10)$$

The initial belief $bel(\mathbf{x}_0)$ must be normally distributed of the form

$$p_{\mathrm{KF}}(\mathbf{x}_0) = \det(2\pi\boldsymbol{\Sigma}_0)^{-\frac{1}{2}} \exp\left(-\frac{1}{2}(\mathbf{x}_0 - \mu_0)^T \boldsymbol{\Sigma}_0^{-1}(\mathbf{x}_0 - \mu_0)\right). \quad (4.11)$$

The linearity assumption for process- and measurement equations (Eq. 4.2 and Eq. 4.3) enables computational efficient matrix multiplication leading to

$$\mathbf{x}_k = \mathbf{F}_k\mathbf{x}_{k-1} + \mathbf{v}_k, \quad (4.12)$$

$$\mathbf{z}_k = \mathbf{H}_k\mathbf{x}_{k-1} + \mathbf{w}_k, \quad (4.13)$$

where $\mathbf{F}_k$ and $\mathbf{H}_k$ represent the process- and measurement matrix and $\mathbf{v}_k$ and $\mathbf{w}_k$ are corresponding zero-mean, uncorrelated and white Gaussian noise terms.

First, the filter procedure predicts the state estimate and corresponding covariance using

$$\hat{\mathbf{x}}_{k-} = \mathbf{F}_k \hat{\mathbf{x}}_{k-1} + \mathbf{B}_k \mathbf{u}_k, \tag{4.14}$$

$$\mathbf{P}_{k-} = \mathbf{F}_k \mathbf{P}_{k-1} \mathbf{F}_k^T + \mathbf{Q}_k, \tag{4.15}$$

where $\hat{\mathbf{x}}_{k-}$ and $\mathbf{P}_{k-}$ are the a priori state estimate and covariance, $\mathbf{B}_k \mathbf{u}_k$ are control input model and control vector describing the system dynamics and $\mathbf{Q}_k$ is the process noise covariance, respectively. For the predicted states an expected measurement and corresponding covariance is calculated with

$$\hat{\mathbf{z}}_{k-} = \mathbf{H}_k \hat{\mathbf{x}}_{k-}, \tag{4.16}$$

$$\mathbf{R}_{k-} = \mathbf{H}_k \mathbf{P}_{k-} \mathbf{H}_k^T. \tag{4.17}$$

Eventually, the observation delivered by the sensor $\mathbf{z} \sim \mathcal{N}(\mathbf{z}, \hat{\mathbf{z}}, \mathbf{R_k})$ at the actual time step k is incorporated in the innovation step determining the innovation γ_k and innovation covariance $\mathbf{S}_k$ with

$$\gamma_k = \hat{\mathbf{z}}_k - \hat{\mathbf{z}}_{k-}, \tag{4.18}$$

$$\mathbf{S}_k = \mathbf{R}_{k-} + \mathbf{R}_k, \tag{4.19}$$

$$\mathbf{S}_k = \mathbf{H}_k \mathbf{P}_{k-} \mathbf{H}_k^T + \mathbf{R}_k. \tag{4.20}$$

Subsequently, the Kalman gain $\mathbf{K}_k$ is determined incorporating the innovation covariance $\mathbf{S}_k$ and the predicted state covariance $\mathbf{P}_{k-}$ in

$$\mathbf{K}_k = \mathbf{P}_{k-} \mathbf{H}_k^T \mathbf{S}_k^{-1}, \tag{4.21}$$

and the state estimate $\hat{\mathbf{x}}_k$ and covariance $\mathbf{P}_k$ is updated using

$$\hat{\mathbf{x}}_k = \hat{\mathbf{x}}_{k-} + \mathbf{K}_k \gamma_k, \tag{4.22}$$

$$\mathbf{P}_k = (\mathbf{I} - \mathbf{K}_k \mathbf{H}_k) \mathbf{P}_{k-}. \tag{4.23}$$

However, in practice, real motions and measurements are rarely of linear kind e.g., a vehicle that moves with a translational and rotational velocity on a circular trajectory. A solution to this problem is the extended Kalman filter (EKF), which relaxes the linearity assumption where the state transition probability and measurement probabilities are governed by nonlinear functions f and h, respectively. Key idea underlying the EKF is the linearization approximating the nonlinear function f by a linear function that is tangent to f at the mean of the posterior $\boldsymbol{\mu}_{k-1}$. Beside many linearization techniques EKF utilizes the first-order Taylor expansion which constructs a linear approximation by the partial derivative with respect to states $\mathbf{x}_k$

$$\boldsymbol{f}'(\mathbf{x}_{k-1}, \mathbf{u}_k) = \frac{\partial \boldsymbol{f}(\mathbf{x}_{k-1}, \mathbf{u}_k)}{\partial \mathbf{x}_{k-1}}, \tag{4.24}$$

$$\boldsymbol{h}'(\mathbf{x}_k) = \frac{\partial \boldsymbol{h}(\mathbf{x}_k)}{\partial \mathbf{x}_k}. \tag{4.25}$$

Hence, $\boldsymbol{f}$ is approximated by its value $\boldsymbol{\mu}_{k-1}$ and the linear extrapolation is achieved by a term proportional to the gradient of $\boldsymbol{f}$ at $\boldsymbol{\mu}_{k-1}$:

$$\boldsymbol{f}(\mathbf{x}_{k-1}, \mathbf{u}_k) = \boldsymbol{f}(\boldsymbol{\mu}_{k-1}, \mathbf{u}_k) + \underbrace{\frac{\partial \boldsymbol{f}(\mathbf{x}_{k-1}, \mathbf{u}_k)}{\partial \mathbf{x}_{k-1}}}_{=:\mathbf{G}_k}(\mathbf{x}_{k-1} - \boldsymbol{\mu}_{k-1}), \tag{4.26}$$

and subsequent for nonlinear measurement function $\boldsymbol{h}$

$$\boldsymbol{h}(\mathbf{x}_k) = \boldsymbol{h}(\bar{\boldsymbol{\mu}}_k) + \underbrace{\frac{\partial \boldsymbol{h}(\mathbf{x}_k)}{\partial \mathbf{x}_k}}_{=:\mathbf{J}_k}(\mathbf{x}_k - \bar{\boldsymbol{\mu}}_k). \tag{4.27}$$

where $\bar{\boldsymbol{\mu}}_k$ are the mean of the a priori state $\hat{\mathbf{x}}_{k-}$. Updated state transition and measurement probabilities, for $p_{\text{EKF}}(\mathbf{x}_k|\mathbf{x}_{k-1})$ and $p_{\text{EKF}}(\mathbf{z}_k|\mathbf{x}_k)$, are

$$p_{\text{EKF}}(\mathbf{x}_k|\mathbf{x}_{k-1}, \mathbf{u}_k) = \det(2\pi \mathbf{Q}_k)^{-\frac{1}{2}}$$
$$\exp\left(-\frac{1}{2}\big(\mathbf{x}_k - \boldsymbol{f}(\mathbf{x}_{k-1}, \mathbf{u}_k)\big)^T \mathbf{Q}_k^{-1}\big(\mathbf{x}_k - \boldsymbol{f}(\mathbf{x}_{k-1}, \mathbf{u}_k)\big)\right), \tag{4.28}$$
$$p_{\text{EKF}}(\mathbf{z}_k|\mathbf{x}_k) = \det(2\pi \mathbf{R}_k)^{-\frac{1}{2}}$$
$$\exp\left(-\frac{1}{2}\big(\mathbf{z}_k - \boldsymbol{h}(\mathbf{x}_k)\big)^T \mathbf{R}_k^{-1}\big(\mathbf{z}_k - \boldsymbol{h}(\mathbf{x}_k)\big)\right), \tag{4.29}$$

The full EKF procedure is then

$$\hat{\mathbf{x}}_{k-} = \boldsymbol{f}(\hat{\mathbf{x}}_{k-1}, \mathbf{u}_k), \tag{4.30}$$

$$\mathbf{P}_{k-} = \mathbf{G}_k \mathbf{P}_{k-1} \mathbf{G}_k^T + \mathbf{Q}_k, \tag{4.31}$$

$$\mathbf{K}_k = \mathbf{P}_{k-} \mathbf{J}_k^T (\mathbf{J}_k \mathbf{P}_{k-} \mathbf{J}_k^T + \mathbf{R}_k)^{-1}, \tag{4.32}$$

$$\hat{\mathbf{x}}_k = \hat{\mathbf{x}}_{k-} + \mathbf{K}_k\big(\hat{\mathbf{z}}_k - \boldsymbol{h}(\hat{\mathbf{x}}_{k-})\big), \tag{4.33}$$

$$\mathbf{P}_k = (\mathbf{I} - \mathbf{K}_k \mathbf{J}_k)\mathbf{P}_{k-}, \tag{4.34}$$

for a single prediction and update iteration.

Practical considerations: Fig. 4.4 shows the Kalman filter flow chart. Note that the flow chart is identical for the EKF, besides that, the state transition and measurement function are of a nonlinear kind and are linearized using a first-order Taylor expansion. The following description is based on [TBF05].

The filter is initialized at $k = 0$ by the mean $\mathbf{x}_0$ and the covariance $\mathbf{P}_0$, e.g., a distance to an object. In case $k \neq 0$, the Kalman filter input is the belief at $k - 1$,

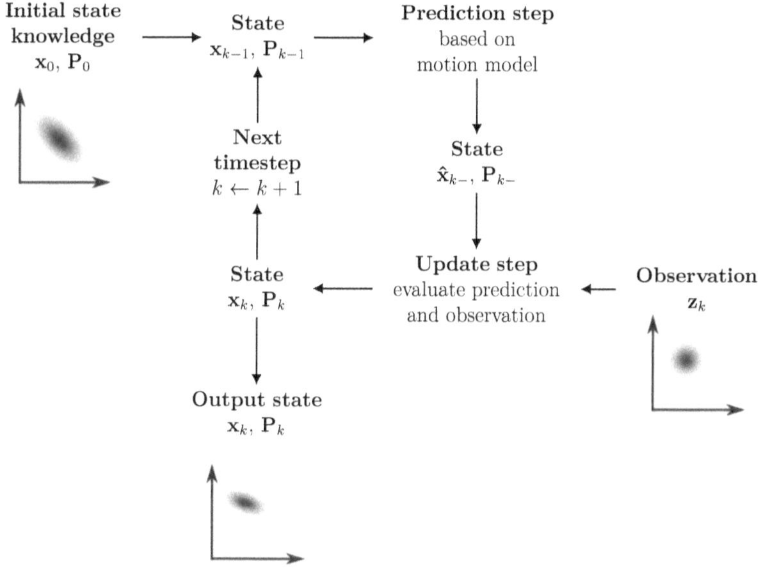

Figure 4.4: Kalman filter procedure.

represented by $\mathbf{x}_{k-1}$ and $\mathbf{P}_{k-1}$. To update these parameters, the filter requires the observation and, if available, a control vector, which provides information about the motion of the ego-system. The output is the belief at time k, represented by $\mathbf{x}_k$ and $\mathbf{P}_k$. The predicted belief $\hat{\mathbf{x}}_{k-}$ and $\mathbf{P}_{k-}$ is calculated based on the belief $\mathbf{x}_{k-1}$ and $\mathbf{P}_{k-1}$ one time step later, but before incorporating the measurement $\mathbf{z}_k$ (Eq. 4.30 and Eq. 4.31).

This belief is obtained by incorporating the control vector, e.g., ego-system accelerations. The mean is updated using the deterministic version of the state transition function, with mean substituted for the state $\mathbf{x}_{k-1}$. The covariance is updated by incorporating the $\mathbf{G}_k$ matrix since states depend on previous states through this matrix. A central parameter in the update step is the Kalman gain $\mathbf{K}_k$ (Eq. 4.32). It specifies the degree to which the measurement $\mathbf{z}_k$ is incorporated into the new state estimate.

Eq. 4.33 manipulates the mean, by adjusting it in proportion to the Kalman gain $\mathbf{K}_k$ and the deviation of the actual measurement $\mathbf{z}_k$ and the measurement predicted according to the measurement probability. The key concept is the *innovation*, which is the difference between the actual measurement $\mathbf{z}_k$ and the expected measurement. Finally, the new covariance of the posterior belief is calculated, adjusting for the information gain resulting from the measurement using Eq. 4.34.

4.2.3 Nonlinear process model

For describing the vehicle motion, pitch and roll rate are neglected. The acceleration a in the vehicle center of mass (COM) is given by

$$a = \dot{v} + \omega \times v,$$

$$= \begin{pmatrix} -v \cdot \sin(\beta)\dot{\beta} \\ v \cdot \cos(\beta)\dot{\beta} \\ 0 \end{pmatrix} + \begin{pmatrix} 0 \\ 0 \\ w \end{pmatrix} \times \begin{pmatrix} v \cdot \cos(\beta) \\ v \cdot \sin(\beta) \\ 0 \end{pmatrix},$$

$$= \begin{pmatrix} -v(w + \dot{\beta})\sin(\beta) \\ v(w + \dot{\beta})\cos(\beta) \\ 0 \end{pmatrix}, \tag{4.35}$$

where β is the side slip angle between vehicle longitudinal direction, $\dot{\beta}$ is the side slip angle velocity, velocity vector v of COM, $\dot{v}$ is the first derivation of velocity and w the yaw rate. This section is based on [KHD$^+$17].

Compared to the original single track model [RS40], an additional force F_{kick} is applied perpendicular to the rear axle. This force is similar to side wind effects [Kam63]. The skid event causes high magnitudes of slip angles α_f and α_r at the front and rear wheels as well as side slip angle β, thus small-angle approximation shall not be employed.

Fig. 4.5 shows the geometry of the single track model. The equilibrium of forces and momentum in lateral vehicle direction with respect to the kick plate force F_{kick} are

$$\dot{w} = \theta^{-1}\Big(F_{y,\text{f}}(\alpha_\text{f}) \cdot l_\text{f} \cdot \cos(\delta) - F_{y,\text{r}}(\alpha_\text{r}) \cdot l_\text{r} + F_{\text{kick}} \cdot l_\text{r}\Big), \tag{4.36}$$

$$\dot{\beta} = \frac{1}{m \cdot v \cdot \cos\beta}\Big(F_{y,\text{f}}(\alpha_\text{f}) \cdot \cos(\delta) + F_{y,\text{r}}(\alpha_\text{r}) - F_{\text{kick}}\Big) - w, \tag{4.37}$$

where δ represents the steering angle of the front wheel and θ the moment of inertia. The mass m of the vehicle is centered in the COM, which has a horizontal distance of l_r from rear axle and l_f from front axle. The lateral tire forces are defined by the product of slip angle $\alpha_{\text{f/r}}$ and corner stiffness $c_{\text{f/r}}$ in linear range

$$F_{y,\text{f/r}} = \alpha_{\text{f/r}} \cdot c_{\text{f/r}}. \tag{4.38}$$

The slip angles $\alpha_{\text{f/r}}$ are defined by the tires velocity vector $v_{\text{f/r}}$. The longitudinal velocities must be equal, since the vehicle is not stretched in longitudinal direction. The difference of lateral velocities is caused by the yaw rate w. Thus, the slip angles $\alpha_{\text{f/r}}$ are defined [Sch14]

$$\alpha_\text{f} = -\arctan\left(\frac{v \cdot \sin(\beta) + l_\text{f} \cdot w}{v \cdot \cos(\beta)}\right) + \delta, \tag{4.39}$$

$$\alpha_\text{r} = -\arctan\left(\frac{v \cdot \sin(\beta) - l_\text{r} \cdot w}{v \cdot \cos(\beta)}\right). \tag{4.40}$$

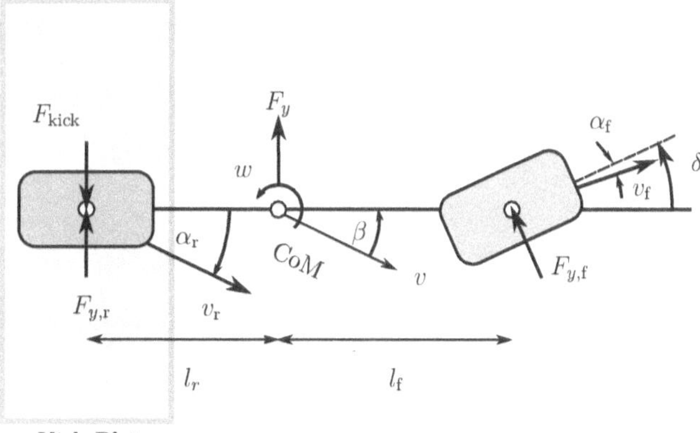

Figure 4.5: Single track motion model sliding towards a target obstacle in a skid scenario. The snapshot shows the force and momentum equilibrium of the model at the application of the external kick plate force F_{kick}. For single track models, the two wheels for front and rear axle are represented by one front wheel and one rear wheel, respectively.

Using Eq. (4.36) − (4.40), a nonlinear model describing the vehicle horizontal motion is derived

$$\dot{w} = \frac{1}{\theta}\Big(c_{\text{f}} \cdot (\alpha_{\text{f}} + \delta) \cdot \cos(\delta) \cdot l_{\text{f}} - c_{\text{r}} \cdot \alpha_{\text{r}} \cdot l_{\text{r}} + F_{\text{kick}} \cdot l_{\text{r}}\Big), \qquad (4.41)$$

$$\dot{\beta} = \frac{1}{mv\cos\beta}\Big(c_{\text{f}} \cdot (\alpha_{\text{f}} + \delta) \cdot \cos(\delta) + c_{\text{r}} \cdot \alpha_{\text{r}} - F_{\text{kick}}\Big) - w, \qquad (4.42)$$

where steering angle δ and kick force F_{kick} are inputs of the model. For skid driving situations the corner stiffnesses $c_{\text{f/r}}$ can not be assumed linear, as shown in Fig. 4.6. The lateral tire force is not proportional to the slip angle. The lateral acceleration a_y is limited to approximately 1 g at favorable conditions and less if the road is irrigated. A solution is to approximate the characteristic lateral forces of the road-tire-contact using the Magic Tyre Formula by Pacejka *et al* [Pac66, Pac05].

$$F_y = D \cdot \sin\Big(C \cdot \arctan\big(B \cdot \alpha - E \cdot (B \cdot \alpha - \arctan(B \cdot \alpha))\big)\Big). \qquad (4.43)$$

The parameter B, C, D and E represent empirical values and tune the curve characteristic. By D, the peak value is defined (for $C \geq 1$) and with curvature factor E, the curve can be additionally stretched or compressed. The factor BCD defines the curve gradient in the coordinate system origin, which is equivalent to corner stiffness at zero slip. The shape factor C scales the curve in x-direction, which is derived from $F_{y,e}$ and D. The parameter B, C, D and E also dominantly depend on the friction coefficient μ and the wheel load.

Fig. 4.6 (a) visualizes the Magic Tyre Formula and Fig. 4.6 (b) shows a model simulation for dry, wet, snow and ice road surface conditions. As expected, the lateral

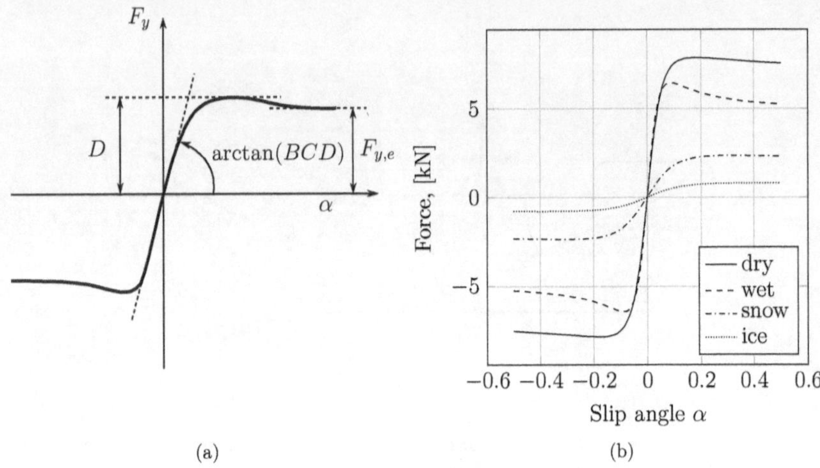

Figure 4.6: Visualization of Magic Tyre Formula by Pacejka *et al* [Pac66, Pac05] (a). Model simulation sweeping slip angle α from -0.5 to 0.5 for dry, wet, snow and ice corresponding model settings for B, C, D, E (b).

tire force is maximal for dry conditions, as the friction coefficient is maximum and decreases to the minimum for icy road conditions. The slip angle is input and an adaptive corner stiffness is output of this model.

4.2.4 Extended Kalman filter implementation

In this work, an extended Kalman filter (EKF) is integrated to estimate the relative position and velocity for a single, static target in skid scenarios. The vehicle motion is determined by incorporating parameters from the proposed modified nonlinear single track motion model considering an additional lateral force F_{kick}. The model is used to determine $\dot{\beta}$ and by integration the side slip angle β serving as an input to the EKF.

Fig. 4.7 shows the geometric relation including a skidding ego vehicle and a present target object. The sensor observations are measured in the coordinate system (SC). Since the target object is assumed static, the relative object position and velocities only depend on the ego vehicle motion. The extended Kalman filter (EKF) system is presented in the following. The system states $\mathbf{x}_k$, the control vector $\mathbf{u}_k$ and the measurement vector $\mathbf{z}_k$ of the vehicle are

$$\mathbf{x}_k = [x_k^{\mathrm{SC}}, \quad y_k^{\mathrm{SC}}, \quad v_{x,k}^{\mathrm{SC}}, \quad v_{y,k}^{\mathrm{SC}}]^{\mathrm{T}}, \tag{4.44}$$

$$\mathbf{u}_k = [a_{x,k}, \quad a_{y,k}, \quad w_k, \quad v_k, \quad \psi_k, \quad \beta_k]^{\mathrm{T}}, \tag{4.45}$$

$$\mathbf{z}_k = [x_k^{\mathrm{SC}}, \quad y_k^{\mathrm{SC}}, \quad v_{x,k}^{\mathrm{SC}}, \quad v_{y,k}^{\mathrm{SC}}]^{\mathrm{T}}, \tag{4.46}$$

where x_k^{SC} and y_k^{SC} represent the relative target object position, $v_{x,k}^{\mathrm{SC}}$ and $v_{y,k}^{\mathrm{SC}}$ the relative target object velocity in the sensor coordinate system.

The ego vehicle motion is estimated by dynamic data provided from an inertial measurement unit (IMU). The measured motion data are the longitudinal and lateral accelerations $a_{x,k}$ and $a_{y,k}$, the yaw rate w_k, by integration of the yaw rate the yaw angle ψ_k, the velocity v_k and the side slip angle (SSA) β from the proposed motion model. The values are stored in the control vector $\mathbf{u}_k$ in each time step k and serve as input to the EKF. The control vector uncertainties are represented by $\mathbf{U}$ with

$$\mathbf{U} = \begin{bmatrix} \sigma_{a_x}^2 & 0 & 0 & 0 & 0 & 0 \\ 0 & \sigma_{a_y}^2 & 0 & 0 & 0 & 0 \\ 0 & 0 & \sigma_w^2 & 0 & 0 & 0 \\ 0 & 0 & 0 & \sigma_v^2 & 0 & 0 \\ 0 & 0 & 0 & 0 & \sigma_\psi^2 & 0 \\ 0 & 0 & 0 & 0 & 0 & \sigma_\beta^2 \end{bmatrix}. \tag{4.47}$$

The ego vehicle position between time step $k-1$ and k is predicted assuming a point mass motion considering the side slip angle which leads to a translation $[\Delta x^e, \Delta y^e]$. Fig. 4.7 shows the ego vehicle motion between two time steps. The translation between two time steps, due to present ego vehicle velocity and yaw rate, is described by

$$\Delta \psi_k = w_k \Delta t, \tag{4.48}$$

$$\Delta x^e = v_k \cos(\beta_k + \Delta \psi_k)\Delta t + \frac{\Delta t^2}{2} a_{x,k}, \tag{4.49}$$

$$\Delta y^e = v_k \sin(\beta_k + \Delta \psi_k)\Delta t - \frac{\Delta t^2}{2} a_{y,k}, \tag{4.50}$$

where $\Delta t = k_1 - k$ represents discrete time step width. The motion to update the position is twofold. First, the instantaneous velocity v_k is split up in $x-$ and $y-$component by considering the total side slip angle (SSA) β and the yaw angle shift since the last time step. The second component incorporates the longitudinal and lateral accelerations. The ego vehicle translation parameter are transformed into the sensor coordinate system using

$$\boldsymbol{\Gamma}^{\mathrm{SC}_{k-1} \to \mathrm{SC}_k} = \begin{bmatrix} \cos(\Delta \psi_k) & -\sin(\Delta \psi_k) & \Delta x^e \\ \sin(\Delta \psi_k) & \cos(\Delta \psi_k) & \Delta y^e \\ 0 & 0 & 1 \end{bmatrix}. \tag{4.51}$$

Accordingly, the position of the object is determined incorporating the previous object position at time step $k-1$ and the translation of the sensor coordinate

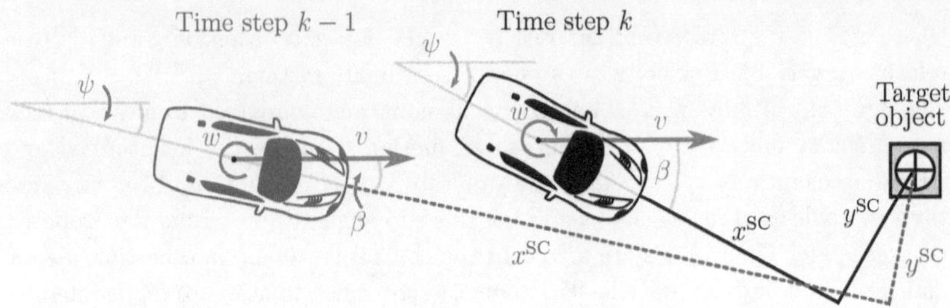

Figure 4.7: Geometric description of instantaneous ego vehicle dynamics for time step $k-1$ and k. The vehicle is skidding towards the target object. The relative target velocity has two components: the instantaneous ego vehicle velocity considering the side slip angle β and the angular velocity due to the yaw rate component.

system at k using

$$
\begin{bmatrix} x \\ y \\ 1 \end{bmatrix}^{SC}_k = \mathbf{\Gamma}^{(SC_{k-1} \to SC_k)^{-1}} \begin{bmatrix} x \\ y \\ 1 \end{bmatrix}^{SC}_{k-1}.
\tag{4.52}
$$

The relative target velocity has two components: the instantaneous ego vehicle velocity (translation) considering the side slip angle β and the angular velocity due to the yaw rate component (rotation). The resulting relative target velocity can be derived using Fig. 4.7 to

$$
\begin{bmatrix} v_x \\ v_y \end{bmatrix}^{SC}_k = \begin{bmatrix} -v_k \cos(\beta_k) + (\sqrt{(x_{k-1}^2 + y_{k-1}^2)} \cdot \sin(\arctan(\frac{y_{k-1}}{x_{k-1}}) - \psi_k) \cdot w_k \\ -v_k \sin(\beta_k) - (\sqrt{(x_{k-1}^2 + y_{k-1}^2)} \cdot \cos(\arctan(\frac{y_{k-1}}{x_{k-1}}) - \psi_k) \cdot w_k \end{bmatrix}.
\tag{4.53}
$$

Hence, the ego motion state transition model can be expressed as

$$
\begin{aligned}
\mathbf{x}_k &= \boldsymbol{f}(\mathbf{x}_{k-1}, \mathbf{u}_k) \\
&= \begin{bmatrix} x_{k-1} \cos(\Delta\psi_k) + y_{k-1} \sin(\Delta\psi_k) - v_k \cos(\beta_k + \Delta\psi_k)\Delta t + \frac{\Delta t^2}{2} a_{x,k} \\ -x_{k-1} \sin(\Delta\psi_k) + y_{k-1} \cos(\Delta\psi_k) - v_k \sin(\beta_k + \Delta\psi_k)\Delta t + \frac{\Delta t^2}{2} a_{y,k} \\ -v_k \cos(\beta_k) + (\sqrt{(x_{k-1}^2 + y_{k-1}^2)} \cdot \sin(\arctan(\frac{y_{k-1}}{x_{k-1}}) - \psi_k) \cdot w_k \\ -v_k \sin(\beta_k) - (\sqrt{(x_{k-1}^2 + y_{k-1}^2)} \cdot \cos(\arctan(\frac{y_{k-1}}{x_{k-1}}) - \psi_k) \cdot w_k \end{bmatrix}^{SC}.
\end{aligned}
\tag{4.54}
$$

As described in section 4.2.2, the EKF uses a first-order Taylor expansion to linearize the state transition model and propagate the state uncertainties $\mathbf{P}_k$. Additionally, the ego motion uncertainties $\mathbf{U}$ are propagated along with the state uncertainties $\mathbf{P}_k$ and represent the total uncertainty of the ego motion:

$$
\mathbf{P}_{k+1} = \mathbf{G}_k \begin{bmatrix} \mathbf{P}_k & 0 \\ 0 & \mathbf{U} \end{bmatrix} \mathbf{G}_k^{T}, \quad \mathbf{G}_k = \frac{\partial \boldsymbol{f}(\mathbf{x}_{k-1}, \mathbf{u}_k)}{\partial(x, y, v_x, v_y, a_x, a_y, w, v, \psi, \beta)}\bigg|_{\hat{\mathbf{x}}_-, \mathbf{u}_k}.
\tag{4.55}
$$

The propagation incorporates the ego motion uncertainty into state space by another
linear approximation considering $\mathbf{U}$ and provides an approximate mapping between
motion noise in control space and motion noise in state space [TBF05].

The measurement vector $\mathbf{z}_k$ provides data of the nearest radar cluster from the
radar cluster list. The determined relative position and velocity serves as input
to the EKF. The measurement uncertainty is described by the covariance $\boldsymbol{R}_k$, is
delivered by the sensor and stored as

$$
\boldsymbol{h}(\hat{\mathbf{x}}) =
\begin{bmatrix}
x & 0 & 0 & 0 \\
0 & y & 0 & 0 \\
0 & 0 & v_x & 0 \\
0 & 0 & 0 & v_y
\end{bmatrix}^{\text{SC}}
, \qquad
\boldsymbol{R}_k =
\begin{bmatrix}
\sigma_x^2 & 0 & 0 & 0 \\
0 & \sigma_y^2 & 0 & 0 \\
0 & 0 & \sigma_{v_x}^2 & 0 \\
0 & 0 & 0 & \sigma_{v_y}^2
\end{bmatrix}
, \qquad (4.56)
$$

and subsequently the Jacobi matrix is

$$
\mathbf{J}_k = \left. \frac{\partial \boldsymbol{h}(\hat{\mathbf{x}})}{\partial(x, y, v_x, v_y)} \right|_{\hat{\mathbf{x}}_{k-}} . \qquad (4.57)
$$

The initial values of the state vector and target position is derived from the latest
measurement before F_{kick} is applied.

4.2.5 Validation in skid scenarios

Kick plates are used to train driver reactions in unexpected skid situations to avoid
over- or understeering and transfer the vehicle back in a safe state [ABBP$^+$11]. In
this work, the kick plate is used to transfer the vehicle, with mounted sensors, in skid
driving situations. Each test run includes three stages: the vehicle is accelerated
to an initial velocity, the rear axle is laterally shifted by the kick plate, causing the
vehicle to increase its yaw angle and side slip angle (SSA) while sliding across the
watered area with low friction. Eventually, the test driver attempts to stabilize the
vehicle while it is skidding towards the target. Fig. 4.1 shows a top view scenario
sequence during one test run taken with a drone.

In regular driving situations, no external force is applied and the test vehicle approaches the EVT straight at velocities between 30 km/h and 50 km/h without significant steering inputs.The kick plate intensity determines the side slip angle of the
test run. The side slip angle is taken as a criticality criterion to distinguish between
regular dynamic scenarios and skid scenarios according to the test method in section
4.1. The SSA defines two groups of test runs: regular and skid scenarios. For regular
test runs, the maximal SSA is smaller than $5°$ and for skid scenarios, the angle is
larger than $5°$, respectively.

In the following, the proposed EKF is validated during a target-tracking task in
an exemplary regular driving and a skid scenario approaching a stationary target.
Fig. 4.8 and Fig. 4.9 present a comprehensive overview about the ego vehicle driving
dynamics (top figure), the achieved state estimation accuracy for position (second-

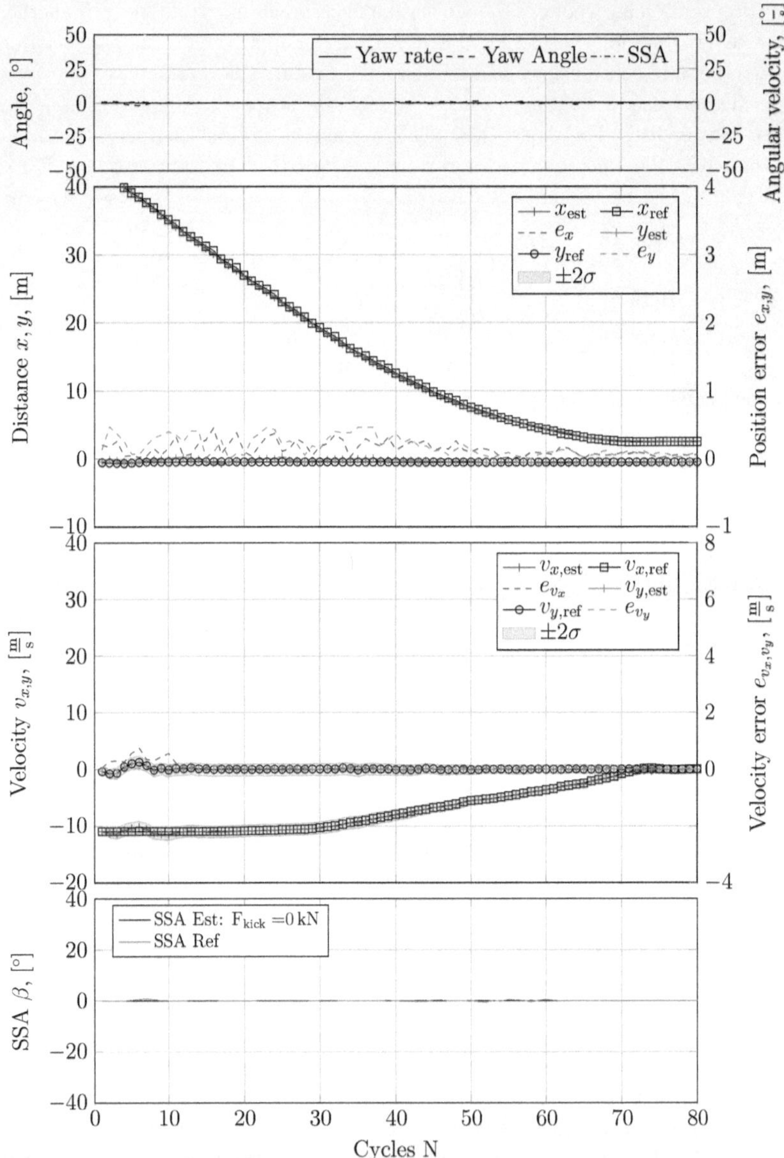

Figure 4.8: EKF results for an exemplary regular driving scenario approaching a static target. Overview about the ego vehicle driving dynamics (top figure), the achieved state estimation accuracy for position (second-row figure) and velocity (third-row figure), the errors when the estimates are related to a reference, the uncertainty represented by $\pm\,2\,\sigma$-bands and the side slip angle estimation (bottom figure) based on the proposed motion model.

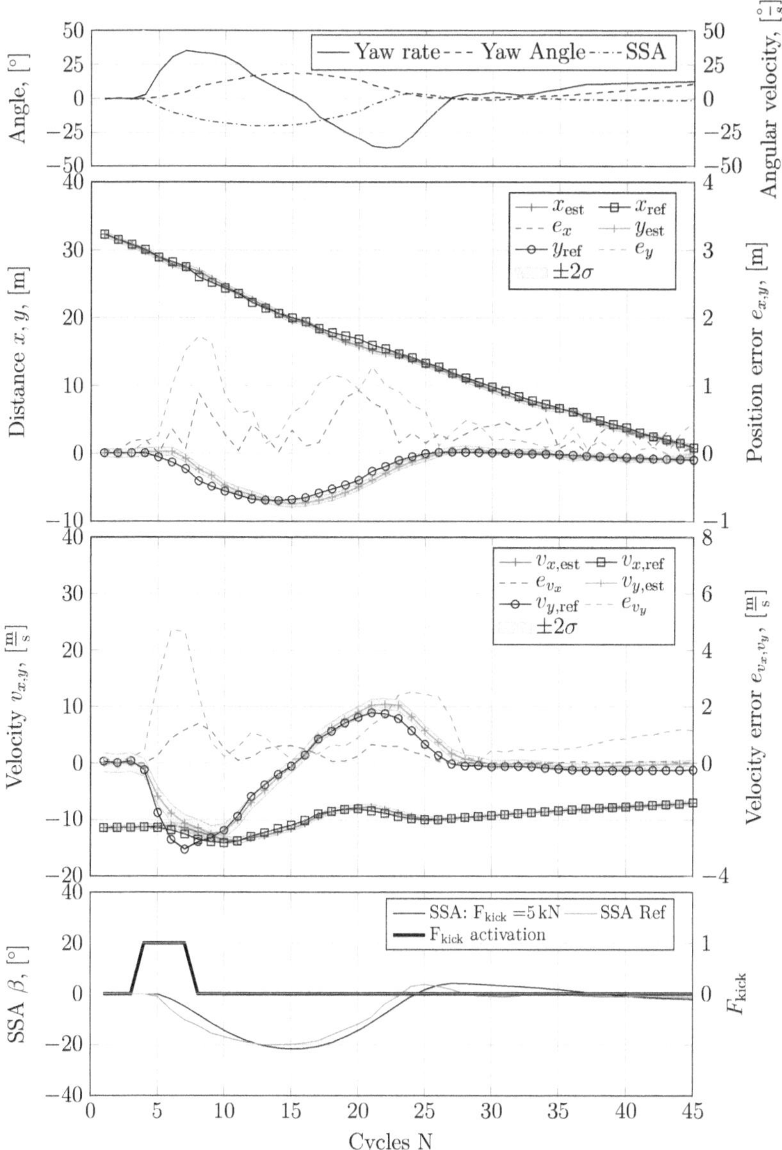

Figure 4.9: EKF results for an exemplary skid scenario sliding towards a static target. Overview about the ego vehicle driving dynamics (top figure), the achieved state estimation accuracy for position (second-row figure) and velocity (third-row figure), the errors when the estimates are related to a reference, the uncertainty represented by $\pm 2\sigma$-bands and the side slip angle estimation (bottom figure) based on the proposed motion model.

row figure) and velocity (third-row figure), the errors when the estimates are related
to a reference, the uncertainty represented by $\pm\,2\,\sigma$-bands and the side slip angle
estimation (bottom plot) based on the proposed motion model, respectively.
During the straight approaching scenario, shown in the top figure of Fig. 4.8, the
yaw angle and rate, as well as the SSA are almost zero. The position and velocity
estimates show a high accuracy compared to the reference sensor. The errors range
below $0.5\,\mathrm{m}$ and $1\,\mathrm{m/s}$ for x-, y-position and v_x-,v_y-velocities, respectively. The bottom plot of Fig. 4.8 shows values close to zero.

Fig. 4.9 presents the EKF performance during a skid scenario sliding towards a
static target. The ego-vehicle rotates counter-clockwise due to the kick plate inducing a maximal yaw rate of $35°/s$, maximal yaw angle of $20°$ and an absolute SSA
in a similar range, as shown in the top plot. To capture this motion the force F_{kick}
is applied to the proposed model for 4 cycles, as shown in the bottom figure. The
number of cycles with additional force is chosen empirically. The ego-vehicle rotation causes a negative y-shift of the target position in the sensor coordinate system
shown in the second-row figure.
The position and velocity estimates show high accuracy compared to the reference
sensor. The maximal position errors are $0.9\,m$ and $1.7\,m$ for x- and y-position, respectively. The maximal x-velocity error is at $1.3\,\mathrm{m/s}$. Immediately after the kick
plate activation, where a high yaw acceleration is present, the lateral velocity error
has its peak with $4.7\,\mathrm{m/s}$ for one cycle and is drastically decreasing afterward.
The bottom figure of Fig. 4.8 shows the side slip angle estimation for several lateral
forces F_{kick} for the skid scenario. The reference side slip angle (red) and the estimated side slip angle using the presented modified motion model with $F_{\mathrm{kick}}= 5\,\mathrm{kN}$
(black) are presented, respectively. The variation of F_{kick} results in different side
slip angles. A common measure to assess tracking quality is to determine the root
mean squeare error (RMSE). This quantity expresses the standard deviation of the
estimation residuals, where a residual is a measure of how far a data point from the
regression line is, and can be expressed as

$$\mathrm{RMSE} = \sqrt{\frac{\sum_{k=1}^{K}(\hat{\mathbf{x}}_k - \mathbf{x}_{REF})^2}{K}}, \tag{4.58}$$

where $\hat{\mathbf{x}}_k$ are the estimates from the proposed EKF and $\mathbf{x}_{REF}$ are the vectorized
reference values obtained from the reference sensor. The RMSE values for both, the
proposed EKF and the series radar, are shown in the following table

Tracking system	Scenario	$\bar{x}$, [m]	$\bar{y}$, [m]	$\bar{v}_x$, [m/s]	$\bar{v}_y$, [m/s]
Proposed EKF	Skid	**0.38**	**0.73**	**0.43**	**1.54**
Series radar	Skid	0.66	1.48	1.46	5.67

Table 4.1: root mean squeare error (RMSE) for proposed EKF and series radar
sensor x, y, v_x, v_y-estimates in one skid test run.

4.2.6 Conclusion

This section emphasizes an extended Kalman filter (EKF)-based target tracker in
skid scenarios, motivated by the insufficient performance of a state-of-the-art automotive radar sensor. First, a test method was developed to evaluate the performance
of a series radar sensor in real skid scenarios. A test vehicle was equipped with a
state-of-the-art automotive radar sensor as the device under test, an IMU as a motion data source and a reference sensor. The sensor system was tested in real skid
situations using the novel test method. The vehicle side slip angle was used as a criticality criterion to classify test runs in regular and skid scenarios. The automotive
radar sensor performance was presented and discussed to emphasize the problem.

As a solution to this problem, the section presents an extended Kalman filter (EKF)
using a modified motion model to estimate the relative position and velocity of a
static object in skid scenarios. The section presents a comprehensive derivation of
the utilized methods and validates the tracking performance with real vehicle tests
carried out on a test site. The proposed EKF significantly increases the accuracy
of the estimated relative parameters compared to an automotive radar sensor. The
motion model of the EKF estimates the side slip angle using the modified motion
model considering an additional lateral force. The model is validated with real vehicle tests in skid driving situations. Based on these results, further development
effort to enable dynamic object tracking and a generic activation strategy seem
appropriate as future work.

4.3 Particle filter based extended object state estimation in pre-crash scenarios

Recent progress in radar development enables high-resolution radar sensors to resolve multiple detections for an extended object. One of the main challenges in radar-based object detection and tracking is the handling of an uncertain measurement model problem dealing with wandering dominant scatter points on the extended object surface, see section 1.3. This phenomenon needs to be considered for close extended object tracking in critical driving situations, e.g., for safety applications. This section presents a SMC tracking framework, also referred to as particle filter (PF), using an extended object measurement model to incorporate multiple wheel hypotheses per target object. The signal processing approach, presented in section 3.2, estimates wheel position and bulk velocities for imminent objects evaluating the micro-Doppler signals. These signals are generated by rotating target wheels, which serve as input to the SMC tracking framework. The presented approach covers the signal processing from raw radar sensor data processing to extended target state estimation. The procedure is evaluated on real experimental data where a target vehicle executes safety-critical dynamic evading maneuvers.

4.3.1 Sequential Monte-Carlo method

The sequential Monte Carlo (SMC), also referred to as particle filter (PF), is an approximate solution to the Bayes recursion, which is applicable to nonlinear non-Gaussian process and observation models. They are an alternative to Gaussian techniques since nonparametric filters do not rely on a fixed functional form of the posterior. Rather they approximate posterior densities by a finite number of values (or samples) representing a region in the state space. For large number of samples nonparametric techniques tend to converge to the correct posterior [TBF05].
The key idea is to represent probability distributions of interest by a set of random samples with associated weights, which is an equivalent representation of the probability density [GSS93]. The SMC method is used for target states estimation due to the high nonlinearity of motion and measurement model. This section presents a brief derivation of the sequential Monte Carlo (SMC) approximation. The notation and principle description is based on [GRA04]. Additional further literature can be found in e.g. [AMGC02, C⁺03, Dau05, DGA00, DDFG01, TBF05].

One considers $N \gg 1$ independent and identically distributed (i.i.d.) samples $\{\mathbf{x}^i\}_{i=1}^{N}$ from an arbitrary probability density $p(\mathbf{x})$. For any function of $g(\mathbf{x})$ the expectation $E_g\{g(\mathbf{x})\}$ can be numerically evaluated by

$$E_g\{g(\mathbf{x})\} = \int g(\mathbf{x})p(\mathbf{x})\,d\mathbf{x} \approx \frac{1}{N}\sum_{i=1}^{N} g(\mathbf{x}^i). \qquad (4.59)$$

The empirical expectation is unbiased and will almost surely converge to the true expectation according to the law of large numbers. The rate of convergence depends on the number of independent samples N, hence the samples can be regarded as a point mass approximation of p, i.e.,

$$p(\mathbf{x}) \approx \frac{1}{N} \sum_{i=1}^{N} \delta(\mathbf{x} - \mathbf{x}^i), \tag{4.60}$$

where δ is the Dirac delta. Commonly, it is not possible to sample efficiently from p. A solution is the application of the importance sampling procedure. N independent and identically distributed (i.i.d.) samples $\{\mathbf{x}^i\}_{i=1}^{N}$ from a known density q are drawn, which is similar to p and referred to as *importance density*. These samples are weighted to receive a weighted point mass approximation to p. The expectation of g can be approximated by the empirical expectation

$$E_g\{g(\mathbf{x})\} = \int g(\mathbf{x})p(\mathbf{x}) \, d\mathbf{x} \approx \frac{1}{N} \sum_{i=1}^{N} w^i g(\mathbf{x}^i), \tag{4.61}$$

with

$$w^i = \frac{\tilde{w}(\mathbf{x}^i)}{\sum_{j=1}^{N} \tilde{w}(\mathbf{x}^j)}, \tag{4.62}$$

$$\tilde{w}(\mathbf{x}^i) = \frac{p(\mathbf{x}^i)}{q(\mathbf{x}^i)}, \tag{4.63}$$

where w^i are the normalized importance density weights and $\tilde{w}$ are the importance weights. For large N, the empirical expectation still converges to the true expectation and hence, we can consider the weighted random samples $\{(w^i, \mathbf{x}^i)\}_{i=1}^{N}$ as point mass approximation of p

$$p(\mathbf{x}) \approx \sum_{i=1}^{N} w^i \delta(\mathbf{x} - \mathbf{x}^i). \tag{4.64}$$

The normalization ensures that all weights sum up to one. The critical operation in particle filtering is to enable a recursive Bayesian state estimation. Therefore, the importance distribution q and weights w^i need to be estimated using the density from the previous time step $k - 1$ and the actual observation at time k. Particle filter methods differ in the generation of the importance distribution q. In this work, the sequential importance resampling (SIR) technique is applied.

The joint posterior density $p(\mathbf{X}_k|\mathbf{Z}_k, \mathbf{u}_k)$ at time k, where $\mathbf{X}_k = \{x_j, j = 0, ..., k\}$ represents the sequence of all target states up to time k, is represented by a set of random samples with associated weights and $\mathbf{u}_k$ is the control vector at time step k. Accordingly, the posterior density $p_k(\mathbf{X}_k|\mathbf{Z}_k)$ at time k is represented as set of

---

Algorithm 1 SIR Particle Filter

1: **procedure** SIR PF($\{\mathbf{x}_{k-1}^i\}_{i=1}^N, \mathbf{u}_k, \mathbf{z}_k$)
2:     **for** $i = 1$ to N **do**
3:         sample $\mathbf{x}_k^i \sim p(\mathbf{x}_k^i|\mathbf{x}_{k-1}^i, \mathbf{u}_k)$
4:         $w_k^i \propto p(\mathbf{z}_k|\mathbf{x}_k^i)$
5:     **end for**
6:     **for** $i = 1$ to N **do**
7:         $\tilde{w}_k^i = (sum[\{w_k^i\}_{i=1}^N])^{-1} \cdot w_k^i$
8:     **end for**
9:     $[\{\mathbf{x}_k^i\}_{i=1}^N]=\textbf{RESAMPLE}[\{\mathbf{x}_k^i, \tilde{w}_k^i\}_{i=1}^N]$
10:     **return** $\{\mathbf{x}_k^i\}_{i=1}^N$
11: **end procedure**

---

weighted particles $\{\mathbf{X}_k^i, w^i\}_{i=1}^N$ and can be approximated as [GRA04]

$$p_k(\mathbf{X}_k|\mathbf{Z}_k) \approx \sum_{i=1}^N w_k^i \delta(\mathbf{X}_k - \mathbf{X}_k^i), \tag{4.65}$$

$$w_k^i = w_{k-1}^i \frac{p(\mathbf{z}_k|\mathbf{x}_k^i)\, p(\mathbf{x}_k^i|\mathbf{x}_{k-1}^i, \mathbf{u}_k)}{q(\mathbf{x}_k^i|\mathbf{X}_{k-1}^i, \mathbf{Z}_k)}. \tag{4.66}$$

This density is not suitable for practical applications, since the path up to $\mathbf{X}_{k-1}^i$, and all observations $\mathbf{Z}_k$ needs to be stored. Therefore, the importance densities dependence is reduced to $\mathbf{x}_{k-1}$ and $\mathbf{z}_k$ that $q(\mathbf{x}_k|\mathbf{X}_{k-1}, \mathbf{Z}_k) = q(\mathbf{x}_k|\mathbf{x}_{k-1}, \mathbf{z}_k)$. The estimate of the marginalized posterior density $p(\mathbf{x}_k|\mathbf{Z}_k)$ and modified weights become

$$p_k(\mathbf{x}_k|\mathbf{Z}_k) \approx \sum_{i=1}^N w_k^i \delta(\mathbf{x}_k - \mathbf{x}_k^i), \tag{4.67}$$

$$w_k^i \propto w_{k-1}^i \frac{p(\mathbf{z}_k|\mathbf{x}_k^i)\, p(\mathbf{x}_k^i|\mathbf{x}_{k-1}^i)}{q(\mathbf{x}_k^i|\mathbf{x}_{k-1}^i, \mathbf{z}_k)}. \tag{4.68}$$

The importance density assumption relaxes the requirement for large data buffers since $q(\mathbf{x}_k|\mathbf{x}_{k-1}, \mathbf{z}_k)$ requires only the state $\mathbf{x}_{k-1}$ from the previous time step $k - 1$ and the actual observation $\mathbf{z}_k$.

The importance weight variance increase over time and has a negative effect on the accuracy [DGA00]. As a consequence, a divergence of the filter due to imbalanced normalized weights after a small number of recursive steps. A solution to overcome the degeneracy problem is to eliminate samples with low importance weights by replacing them with samples that have high importance weights so that the random measures $\{(\mathbf{x}_k^i, w_k^i)\}$ are replaced by $\{(\mathbf{x}_k^{i*}, 1/N)\}$. One suitable method to achieve the replacement is called the sequential importance resampling (SIR) technique.

Therefore, the transitional prior $p(\mathbf{x}_k|\mathbf{x}_{k-1}^i)$ with zero-mean Gaussian process noise is used as importance density. Since we apply a resampling step at every time index and set the importance weights to $w_{k-1}^i = 1/N$ for all $i = 1, ..., N$ there is no need to pass on the importance weights from one time step to the next. The relationship

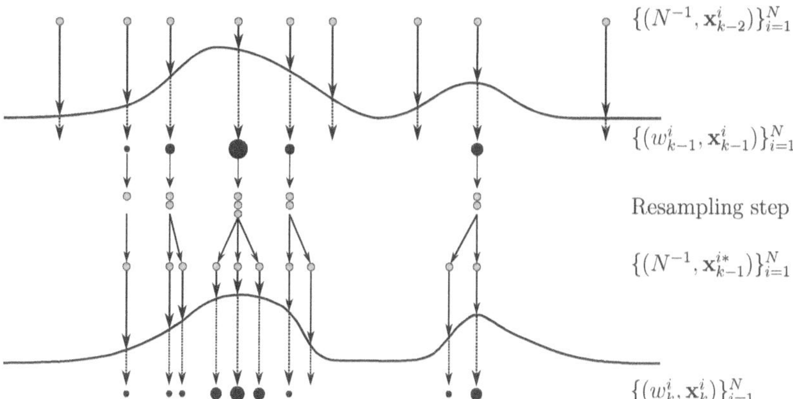

Figure 4.10: Representation of the sequential importance resampling (SIR) algorithm of the particle filter.

(4.68) simplifies to

$$w_k^i \propto p(\mathbf{z}_k | \mathbf{x}_k^i). \tag{4.69}$$

The sequential importance resampling (SIR) particle filter predicts samples from previous time step $k-1$ using the process model representing the system with additional process noise followed by a subsequent approximation based on an evaluation of the measurement likelihood.

The full sequential importance resampling (SIR) particle filter routine for one iteration is shown in algorithm 1. There are various resampling subroutines in the literature, but for this work, the low variance sampling is used.

As a brief overview, the basic idea is to decide which particle is replaced involves another stochastic process [TBF05]. This resampling algorithm draws a scalar sample from a Gaussian distribution and selects particles according to this number. These particles exhibit still a probability proportional to the sample weight. This is achieved by drawing a random number s in the interval $[0; N^{-1}]$ and the algorithm iteratively selects particles by adding the fixed amount M^{-1} to the random number s and by choosing the particle that corresponds to the resulting number. The particle i, for which the following expression holds true is selected

$$i = \arg \min_{N} \sum_{m=1}^{N} \tilde{w}_k^{[m]} \geq (s + (m-1) \cdot M^{-1}). \tag{4.70}$$

Practical considerations: Fig. 4.10 shows an exemplary particle filter procedure sequence. A set of particles $\{N^{-1}, \mathbf{x}_{k-2}^{i}\}_{i=1}^{N}$ starts at time $k = -2$. The particles provide an approximation of the initial probability distribution $p(\mathbf{x}_{k-2})$. In the next step, the importance weights w_{k-1}^{i} are computed using the information available at time $k-1$, e.g., a set of observations. This results in the weighted measure $\{(w_{k-1}^{i}, \mathbf{x}_{k-1}^{i})\}_{i=1}^{N}$ yielding an approximation $p(\mathbf{x}_{k-1} | \mathbf{Z}_{k-1})$. Subsequently, the resampling step replaces particles with a small weights by copies of the fittest particles to obtain the unweighted measure $\{(N^{-1}, \mathbf{x}_{k-1}^{i*})\}_{i=1}^{N}$. The resampling is accomplished using the sequential importance resampling technique, which uses Eq. 4.70. Finally, the prediction step introduces variety, resulting in the measure $\{(N^{-1}, \mathbf{x}_{k-1}^{i*})\}_{i=1}^{N}$, which is an approximation of $p(\mathbf{x}_k|\mathbf{Z}_{k-1})$ [DDFG01].

4.3.2 Related work

A current state of the art overview for extended object tracking is given in [GBR16]. Besides giving an overview of popular tracking procedures, the authors cluster the main measurement likelihood methods in groups. The first group assumes a set of points on a rigid body (SPRB) with a deterministic number of reflection points fixed on a rigid body shape, which was applied to automotive radar data in [BY06, HLS12, HSSS12, GSDB07]. The second method for extended object tracking defines a spatial distribution to model the target detections around the target. The first approach proposed by Gilholm et al. [GGMS05] assumes the number of detections as Poisson distribution. The third paradigm models the measurement likelihood based on physical properties. A direct scattering extended object model is presented describing the radar measurements like range and Doppler shift as several probability distributions [KSD16]. The last method is based on machine learning, e.g., a variational Gaussian mixture measurement model is developed where each mixture component can be interpreted as a particular reflection center [SD18].

Another approach is to incorporate all available measurements using extended object measurement methods and Bayesian state estimation techniques under certain assumptions, e.g., where a radar model considering actual radar sensor resolution is taken into account [HSSS12] or an assumed elliptical object shape [RS04, SP03].

A spatial model-based approach of an extended object was presented by Knill *et al.* [KSD16]. A direct scattering model for high-resolution radars is used to track single extended vehicles. The model assumes vehicles as boxes with an approximately rectangular shape and estimates the joint measurement likelihood for every detection. The measurement likelihood is calculated considering individual likelihoods for range, velocity and angle for each detection. The model is deployed in a Rao-Blackwellized particle filter (RBPF) dividing the state vector in a nonlinear subroutine estimating the target motion and a linear subroutine estimating the objects geometric.

Kellner *et al.* [KBK+16] presents an approach to incorporate multiple radar reflections from an extended object where the authors take the Doppler velocity distribution across the entire object into account. The key assumption is to eliminate the target object rotational proportion of the radial velocity. The object velocity vector of interest is derived using a regression method for multiple detections backscattered from the extended object. The object tracking is realized using a KF with different observation models to evaluate tracking performance.

In contrast, Yang *et al.* [YGMB07] analyzed the range-Doppler profile of an extended target. The authors developed a maneuver indicator based on the rotational velocity component of the radial velocity, which leads to multiple Doppler velocities in a single range cell.

Another method to model the target object is to apply finite set statistics. In this concept, the representation of present targets, measurement detections, motion models and measurement models are modeled as random variables summarized in random sets. A introduction is given in [GLGO14, Mah07]. The posterior density over this multi-object state is then estimated recursively in a similar fashion to the standard Bayes filter [Mah07]. Since both, the object states and the measurements, are modeled as RFS, association ambiguities and clutter can be treated probabilistically and are filtered over time.

A RFS approach based on labeled multi-bernoulli (LMB) filter in multi-sensor multi-object traffic scenarios is presented by Scheel *et al.*, where the object states, measurements, clutter and missed detections are modeled as RFS [SKRD16]. The measurement model from [KSD16] is used to fuse target-level observations from two radar sensors. Cluttered measurements originated from spinning wheels next to static obstacles and close to road borders are stated as a significant difficulty for the association procedure.

Scheel *et al.* [SD18] presents a variational radar model for tracking vehicles using radar detections which are processed by a RFS-based filter. The measurement model is a variational Gaussian mixture approach trained using real radar data, where the authors include measurement positions deviation between the expected Doppler velocity and the measured Doppler velocity. It is used to determine a predictive density for radar detections given a particular state.

In this work, a RFS-based extended object measurement model estimating the measurement likelihood to incorporate the preprocessed wheel hypotheses deployed in a particle filter environment.

4.3.3 Process model

This section presents the target motion model evolution over time, discusses the state parameterization and discretized motion model.

Successful target tracking requires an appropriate motion model representing the target state evolution over time and should further perform highly accurate in diverse driving scenarios. During the last decades, various mathematical models to describe the target motion have been developed and reported in the literature, such as [LJ03, BP99, BSWT11]. The challenge for object state estimation is to achieve a balance between tracking accuracy and the computational complexity of the motion model. Less sophisticated models, e.g., constant velocity (CV) or constant acceleration (CA), have the main advantage in their linearity, allowing an optimal propagation of the state probability distribution but do not consider rotations, e.g., the yaw rate.

More sophisticated models, e.g., the constant turn rate and velocity (CTRV) model or constant turn rate and acceleration (CTRA) model, also referred to as curvilinear

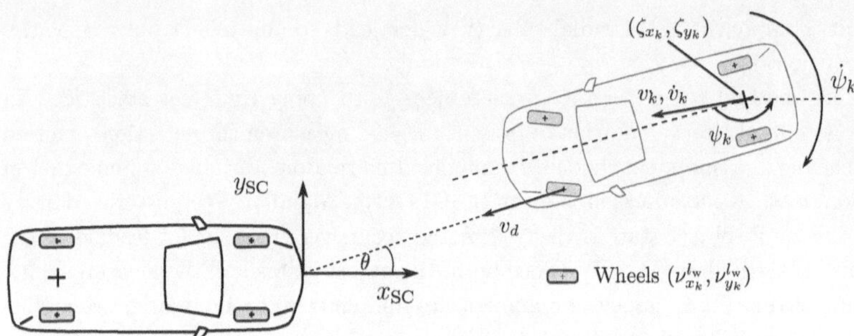

Figure 4.11: The target relative position and dynamics is estimated for its center of motion. The measurements model is based on up to four extracted wheels from the raw radar data.

models, take rotations around the z-axis into account. Overall, the latter curvilinear models describe the motion of road vehicles in low dynamic scenarios sufficiently accurate. To model the motion in high dynamic scenarios, e.g., drifting or skidding, the side slip angle and varying nonlinear tire forces need to be taken into account. However, Schubert *et al.* [SRW08] evaluates the CV, CTRV, CTRA and constant curvature and acceleration (CCA) model by incorporating the motion models along with global positioning system (GPS) data in several unscented Kalman filter (UKF) and evaluated the results with the GPS reference data. It has been shown, that the CTRA model exhibit more accurate performance than the CTRV model and at the same time, is computational less complex than the CCA model.

Hence, in this work, the CTRA model [GSDB07] is used to determine the target states

$$\boldsymbol{\xi}_k = [\zeta_{x_k}, \ \zeta_{y_k}, \ \psi_k, \ v_k, \ w_k, \ \dot{v}_k]^T, \tag{4.71}$$

where ζ_{x_k} and ζ_{y_k} are the target motion center (center of rear axle) coordinates in the sensor coordinate system, ψ_k is the heading angle, v_k is the velocity in the heading direction of the object, w_k is the yaw rate and $\dot{v}_k$ is the acceleration at time k respectively, see Fig. 4.11. The states propagate over time according to the discretized state transition model

$$\boldsymbol{\xi}_k = \boldsymbol{\xi}_{k-1} + \begin{pmatrix} \Delta\zeta_{x_k} \\ \Delta\zeta_{y_k} \\ \dot{\psi}_{k-1}\, T \\ \dot{v}_{k-1}\, T \\ 0 \\ 0 \end{pmatrix} + \begin{pmatrix} 0 \\ 0 \\ \frac{T^2}{2}\, e_{\ddot{\psi}_{k-1}} \\ \frac{T^2}{2}\, e_{\ddot{v}_{k-1}} \\ T\, e_{\ddot{\psi}_{k-1}} \\ T\, e_{\ddot{v}_{k-1}} \end{pmatrix}, \tag{4.72}$$

with

$$\Delta\zeta_{x_k} = \zeta_{x_k} - \zeta_{x_{k-1}} = \cos(\psi_{k-1}) \, K_1 - \sin(\psi_{k-1}) \, K_2, \qquad (4.73)$$

$$\Delta\zeta_{y_k} = \zeta_{y_k} - \zeta_{y_{k-1}} = \sin(\psi_{k-1}) \, K_1 + \cos(\psi_{k-1}) \, K_2, \qquad (4.74)$$

and

$$K_1 = v_{k-1} \, T + \frac{\dot{v}_{k-1} \, T^2}{2} + \frac{e_{\ddot{v}_{k-1}} \, T^3}{6}, \qquad (4.75)$$

$$K_2 = \frac{v_{k-1}\dot{\psi}_{k-1} \, T^2}{2} + \frac{(v_{k-1}e_{\ddot{\psi}_{k-1}} + 2\dot{v}_{k-1}\dot{\psi}) \, T^3}{6}$$

$$+ \frac{(e_{\ddot{v}_{k-1}}\dot{\psi} + \dot{v}_{k-1}e_{\ddot{\psi}_{k-1}}) \, T^4}{8} + \frac{e_{\ddot{v}_{k-1}}e_{\ddot{\psi}_{k-1}} \, T^5}{20}, \qquad (4.76)$$

where noise terms are constant for a single time step T with $T = t_k - t_{k-1}$, and $e_{\ddot{\psi}_k}$, $e_{\ddot{v}_k}$ are white zero-mean Gaussian noise processes with variances $\sigma^2_{\ddot{\psi}}$ and $\sigma^2_{\ddot{v}}$, respectively.

4.3.4 Particle filter implementation

Fig. 4.11 introduces the coordinate systems required for this approach. The raw data signal processing, presented in section 3.2.1, to detect the micro-Doppler based features and obtain characteristic wheel hypotheses, is applied to every radar measurement snapshot. The measurement model incorporates the position and radial velocity information for each wheel hypothesis.

A brief definition for random finite set (RFS), according to [Mah07], is given as: Assuming a random variable Υ that draws its instantiations $\Upsilon = \boldsymbol{Y}$ from the hyperspace Υ_0 of all finite subsets $\boldsymbol{Y}$ (the null set $\emptyset$ included) of some underlying space Υ_0. In this case, the underlying space Υ_0 is a Euclidean vector space, e.g., $\Upsilon_0 = \mathbb{R}^{n_y}$. The approach exploits the RFS property, that the measurement likelihood $p(\mathbf{z}|\mathbf{x})$ incorporating the wheel hypotheses, is capable of processing varying numbers of detected wheels or an empty set.

The general measurement model is briefly presented in the following and based on [Mah07]. A single sensor with state $\mathbf{x}^*$ and the following likelihood function

$$f_{k+1}(\mathbf{z}|\mathbf{x}) \triangleq f_{W_{k+1}}(\mathbf{z} - \eta_{k+1}(\mathbf{x}, \mathbf{x}^*)), \qquad (4.77)$$

is observing a scene. The following probabilities for detections generated by a target, a missed detection and clutter are introduced:

- probability $\boldsymbol{f}_k(\mathbf{z}|\mathbf{x})$, the target will generate observations $\mathbf{z}$ if it has state $\mathbf{x}$, referred to as $p_D(\mathbf{x}, \mathbf{x}^*)$.

- probability $p_D(\mathbf{x})$, the target will not generate an observation, referred to as missed detection with probability $1 - p_D(\mathbf{x})$.

- probability $c_{k|k}(\mathbf{Z})$, a set $\mathbf{Z} = \{\mathbf{z_1}, ..., \mathbf{z_m}\}$ of clutter observations will be generated and assumed as Poisson distributed $c_{k|k}(\mathbf{Z}) = e^{\lambda} \lambda_{k|k}^{n} c_{k|k}(\mathbf{z_1}) \ldots c_{k|k}(\mathbf{z_m})$, where observations and clutter are independent.

Based on these assumption the true likelihood function $f_{k+1}(\mathbf{Z}|\mathbf{X})$ for point targets can be derived to

$$f_{k+1}(\emptyset|\mathbf{X}) = e^{-\lambda} \prod_{\mathbf{x} \in \mathbf{X}} (1 - p_D(\mathbf{x})), \tag{4.78}$$

$$f_{k+1}(\mathbf{Z}|\mathbf{X}) = e^{\lambda} f_C(\mathbf{Z}) \cdot f_{k+1}(\emptyset|\mathbf{X})$$
$$\sum_{\theta} \prod_{i:\theta(i)>0} \frac{p_D(\mathbf{x}_i) \cdot f_{k+1}(\mathbf{z}_{\vartheta(i)}|\mathbf{x}_i)}{(1 - p_D(\mathbf{x}_i)) \cdot \lambda c(\mathbf{z}_{\vartheta,(i)})}, \tag{4.79}$$

where the summation is taken over all association hypotheses ϑ, $\mathbf{X} = \mathbf{x}_1, \ldots, \mathbf{x}_n$, $\mathbf{Z} = \mathbf{z}_1, \ldots, \mathbf{z}_m$ and

$$f_{C(X)}(\mathbf{Z}) = e^{-\lambda(\mathbf{X})} \prod_{\mathbf{z} \in \mathbf{Z}} \lambda \cdot c(\mathbf{z}). \tag{4.80}$$

Eq. (4.78) and Eq. (4.79) represent the true likelihood for point targets.

In the following, this standard model is extended to a single extended target likelihood model to incorporate multiple observations from the target per iteration [Mah07]. The true measurement likelihood for a single extended object is modeled as RFS. Given the sensor with state $\mathbf{x}^*$, probability of detection $p_D(\mathbf{x}, \mathbf{x}^*)$ and likelihood function $f_{k+1}(\mathbf{z}|\mathbf{x}) \triangleq f_{W_{k+1}}(\mathbf{z} - \eta_{k+1}(\mathbf{x}, \mathbf{x}^*))$. As presented in section 4.3.3, the general state of the extended target is

$$\boldsymbol{\xi} = [\zeta_x, \ \zeta_y, \ \psi, \ v, \ w, \ \dot{v}]^T. \tag{4.81}$$

The extended target is modeled as a collection of point scatterers, representing the wheel positions and velocity, fixed on the target surface

$$\check{\mathbf{x}}^1 + \mathbf{x} \quad , ..., \check{\mathbf{x}}^{L_{\mathrm{w}}} + \mathbf{x}. \tag{4.82}$$

The scatter points $\check{\mathbf{x}}^1, ..., \check{\mathbf{x}}^{L_{\mathrm{w}}}$ incorporate the wheel hypotheses and, therefore have the form

$$\mathbf{x}^{l_{\mathrm{w}}} = [x_{l_{\mathrm{w}}}, \ y_{l_{\mathrm{w}}}, \ 0, \ v_{d,l_{\mathrm{w}}}, \ 0, \ 0]^T. \tag{4.83}$$

The detection probability $p_D^{l_{\mathrm{w}}}(\mathbf{x}, \mathbf{x}^*)$ depends on the visibility function $e^{l_{\mathrm{w}}}(\mathbf{x}, \mathbf{x}^*)$, which describes the occlusion state of the scatter point, e.g., one wheel is occluded by another wheel and therefore unlikely to be visible to the sensor, by

$$e^{l_{\mathrm{w}}}(\mathbf{x}, \mathbf{x}^*) = \left\{ \begin{array}{ll} 1 & \text{if} \quad \text{scatterer point is in LoS} \\ 0 & \text{if} \quad \quad \quad \text{otherwise} \end{array} \right\}. \tag{4.84}$$

The probability of detection of point scatter $\breve{\mathbf{x}}^{l_w} + \mathbf{x}$ is then

$$p_D^{l_w}(\mathbf{x}, \mathbf{x}^*) = e^{l_w}(\mathbf{x}, \mathbf{x}^*) \cdot p_D(\breve{\mathbf{x}}^{l_w} + \mathbf{x}, \mathbf{x}^*). \tag{4.85}$$

Likewise, the deterministic measurement model at the point scatterer $\breve{\mathbf{x}}^{l_w} + \mathbf{x}$

$$\eta^{l_w}(\mathbf{x}, \mathbf{x}^*) \triangleq \eta_{k+1}(\breve{\mathbf{x}}^{l_w} + \mathbf{x}, \mathbf{x}^*), \tag{4.86}$$

and the corresponding likelihood function is

$$f_{k+1}^{l_w}(\mathbf{z}|\mathbf{x}) \triangleq f_{\mathbf{W}_{k+1}}(\mathbf{z} - \eta^{l_w}(\mathbf{x}, \mathbf{x}^*)). \tag{4.87}$$

In case there is no measurement assigned to a point scatterer the empty set is defined as

$$\emptyset^{l_w}(\mathbf{x}, \mathbf{x}^*) = \emptyset^{p_D^{l_w}(\mathbf{x}, \mathbf{x}^*)}. \tag{4.88}$$

where $\emptyset^{l_w}(\mathbf{x}, \mathbf{x}^*)$ is the random subset of hyperspace Υ_0. Hence,

$$\Pr(\emptyset^{l_w}(\mathbf{x}, \mathbf{x}^*) = T) \triangleq \begin{cases} p_D^{l_w}(\mathbf{x}, \mathbf{x}^*) & \text{if} \quad T = \Upsilon_0 \\ 1 - p_D^{l_w}(\mathbf{x}, \mathbf{x}^*) & \text{if} \quad T = \emptyset \\ 0 & \text{if} \quad \text{otherwise} \end{cases}. \tag{4.89}$$

The random set measurement model at site $\breve{\mathbf{x}}^{l_w} + \mathbf{x}$ is

$$\Upsilon^{l_w}(\mathbf{x}, \mathbf{x}^*) = \{\eta^{l_w}(\mathbf{x}, \mathbf{x}^*) + \mathbf{W}_{l_w}\} \cup \emptyset^{l_w}(\mathbf{x}, \mathbf{x}^*) \tag{4.90}$$

for all point scatterers, where $\mathbf{W}_1, ..., \mathbf{W}_{l_w}$ are i.i.d. random vectors with density function $f_{\mathbf{W}_{k+1}}(\mathbf{z})$. The measurement model over all point scatterers is thus

$$\Sigma_{k+1} = \Upsilon(\mathbf{x}) \cup \boldsymbol{C}, \tag{4.91}$$

where $\Upsilon(\mathbf{x})$ are detections generated by the target, $\boldsymbol{C}$ represent the false detections and $\Upsilon(\mathbf{x}) = \Upsilon^1(\mathbf{x}) \cup ... \cup \Upsilon^{L_w}(\mathbf{x})$. Therefore one can extend the belief-mass functions at a single point scatter from using Eq. (4.78) and Eq. (4.79) and

- $p_D^{l_w}(\mathbf{x}, \mathbf{x}^*)$ for $p_D(\mathbf{x}_i, \mathbf{x}^*)$,

- $\eta^{l_w}(\mathbf{x}, \mathbf{x}^*)$ for $\eta_{k+1}(\mathbf{x}_i, \mathbf{x}^*)$

- $f_{k+1}^{l_w}(\mathbf{z}, \mathbf{x}^*)$ for $\boldsymbol{f}_{k+1}(\mathbf{z}|\mathbf{x}_i)$

for $\mathbf{Z} \neq \emptyset$ to

$$\boldsymbol{f}_{k+1}(\mathbf{Z}|\mathbf{x}) = e^{\lambda} f_C(\mathbf{Z}) \cdot \boldsymbol{f}_{k+1}(\emptyset|\mathbf{x}) \sum_{\Theta} \prod_{\Theta(l_w)>0} \frac{p_D^{l_w}(\mathbf{x}) \cdot \boldsymbol{f}_{k+1}^{l_w}(\mathbf{z}_{\Theta(l_w)}|\mathbf{x})}{(1 - p_D^{l_w}(\mathbf{x})) \cdot \lambda c(\mathbf{z}_{\vartheta(l_w)})}, \tag{4.92}$$

and for $\mathbf{Z} = \emptyset$

$$\boldsymbol{f}_{k+1}(\emptyset|\mathbf{x}) = e^{-\lambda} \prod_{l_{\mathrm{w}}=1}^{L_{\mathrm{w}}} (1 - p_D^{l_{\mathrm{w}}}(\mathbf{x})). \tag{4.93}$$

For this **book**, the clutter term $f_C(\mathbf{Z})$ is neglected. The clusters possess predefined positions on the structure, represented by $\{\nu_{x,k}^{l_{\mathrm{w}}}, \nu_{y,k}^{l_{\mathrm{w}}}, v_d^{l_{\mathrm{w}}}\}$ where $l_{\mathrm{w}} = 1, ..., L_{\mathrm{w}}$ are the detected wheels and $\nu_{x,k}^{l_{\mathrm{w}}}$, $\nu_{y,k}^{l_{\mathrm{w}}}$ and $v_d^{l_{\mathrm{w}}}$ are the corresponding x- and y-coordinates and the radial velocity component for each hypothesis, respectively. Accordingly, the measurement vector for $\mathbf{Z} \neq \emptyset$ at time step k is

$$\mathbf{z}_k = \left[\nu_{x,k}^1, \nu_{y,k}^1, v_{d,k}^1, \ldots, \nu_{x,k}^{L_{\mathrm{w}}}, \nu_{y,k}^{L_{\mathrm{w}}}, v_{d,k}^{L_{\mathrm{w}}}\right]^T. \tag{4.94}$$

The deterministic measurement model $\eta^{l_{\mathrm{w}}}(\mathbf{x}, \mathbf{x}^*)$ is composed of three individual components transcribing the object position and velocity in measurement space. The first and second component describes the conversion from polar sensor coordinate system to the Cartesian vehicle coordinate system from the wheel clusters. The third component determines the object expected Doppler velocity for every wheel hypothesis, which can be measured by the sensor.

The velocity vector of any given point on the extended object is independent of range and described through the radial proportion of the superposition of the object translational and rotational velocity [KSD16]

$$v_d^{l_{\mathrm{w}}}(\theta, w, \psi) = v\cos(\theta - \psi) + w(\nu_y \cos(\theta) - \nu_x \sin(\theta)). \tag{4.95}$$

The term $\boldsymbol{f}_{k+1}^{l_{\mathrm{w}}}(\mathbf{z}_{\Theta(l_{\mathrm{w}})}|\mathbf{x})$ describes the measurement probability for a given state. The probabilities are determined using the predicted states and the observations as wheel hypotheses delivered by the sensor and the pre-processing stage as Gaussian distributions

$$\mathcal{N}(\mathbf{z}_{\Theta(l_{\mathrm{w}})}, \mathbf{x}, \boldsymbol{R}_{\mathbf{z}_k^{l_{\mathrm{w}}}}), \tag{4.96}$$

where measurement uncertainty for a single wheel is set as white Gaussian noise described by the covariance $\boldsymbol{R}_{\mathbf{z}_k^{l_{\mathrm{w}}}}$ with

$$\boldsymbol{R}_{\mathbf{z}_k^{l_{\mathrm{w}}}} = \begin{bmatrix} \sigma_x^2 & 0 & 0 \\ 0 & \sigma_y^2 & 0 \\ 0 & 0 & \sigma_{v_d}^2 \end{bmatrix}. \tag{4.97}$$

Eq. (4.94)-(4.97) parameterize the measurement likelihood Eq. (4.92). Generally speaking, it is a measure of how the predicted states and the measurement in measurement space resemble. The data association for single extended object generated radar detection is coped using the simple yet effective global nearest neighbor approach with a minimum threshold. The particle filter implementation utilizes the motion model, presented in section 4.3.3, the measurement model, presented in this section and the prediction and update procedure presented in section 4.3.1.

4.3.5   Validation in pre-crash scenarios

This section presents an evaluation of the proposed SMC-based tracking methods. The tracking results are evaluated using the identical measurement setup, as presented in section 3.2.2, enabling comprehensive parameter validation. Again, radar sensor measurements were conducted using a real vehicle as target object equipped with a RTK reference measurement system to evaluate the presented methods. The target vehicle is equipped with a GeneSys ADMA-G-Pro+ with RTK positioning to obtain the ground truth global position estimate with an accuracy of $\pm 0.02\,\mathrm{m}$ and acceleration and turn rate data in three-axis for arbitrary positions of the target vehicle.

A high dynamic evading maneuver in front of the radar sensor is the evaluation scenario. These test runs are safety-critical when the test driver attempts to avoid an imminent collision at the very last moment. Fig. 3.7 describes the critical evading maneuver and its driving dynamics. The top figure visualizes the wheel positions of the target vehicle. The center and bottom figures show the target velocity and yaw rate. The actual dynamic target parameters are presented in Fig. 4.12 and Fig. 4.13. The dashed graphs in each subfigure show the true target states.
SMC implementations generate and propagate particles in a random manner, where the state estimation results may be distorted. The results are averaged over 20 runs to minimize these effects [SD18]. The dimensions of the deterministic measurement likelihood model for x- and y-position from the wheels was matched to the target vehicle. The values for target wheelbase and tread are set to $3\,\mathrm{m}$ and $2\,\mathrm{m}$, respectively. The target length and width to evaluate detections are set to $5\,\mathrm{m}$ and $2\,\mathrm{m}$.

Fig. 4.12 shows the estimates x-, y-position and velocity compared to the true target states measured with a RTK reference system. The target vehicle approaches the sensor with $11\,\mathrm{m/s}$, which leads to a collision if the test driver does not execute an evading maneuver to avoid the collision.
The top figure presents the number of detected wheel hypotheses. The subsequent figures present the x-, y- and v estimates (solid lines), the true target motion (dashed lines) and $\pm 2\sigma$ standard deviation of the estimate for each cycle. An error plot for each parameter is shown. The error values for x and y state estimates range between $-0.2\,\mathrm{m}$ and $0.4\,\mathrm{m}$. The velocity error has its maximum at $0.3\,\mathrm{m/s}$.
The wheel detection procedure detects a minimum of two wheels up to cycle 41, which leads to an overall accurate target state estimation. The resulting accuracy is as expected since the target is with its full extent in the sensor FoV. From time step 42, the number of detected wheel hypotheses decreases to one, which leads to increased uncertainty for the measurement model, and hence, to a decreased tracking performance.

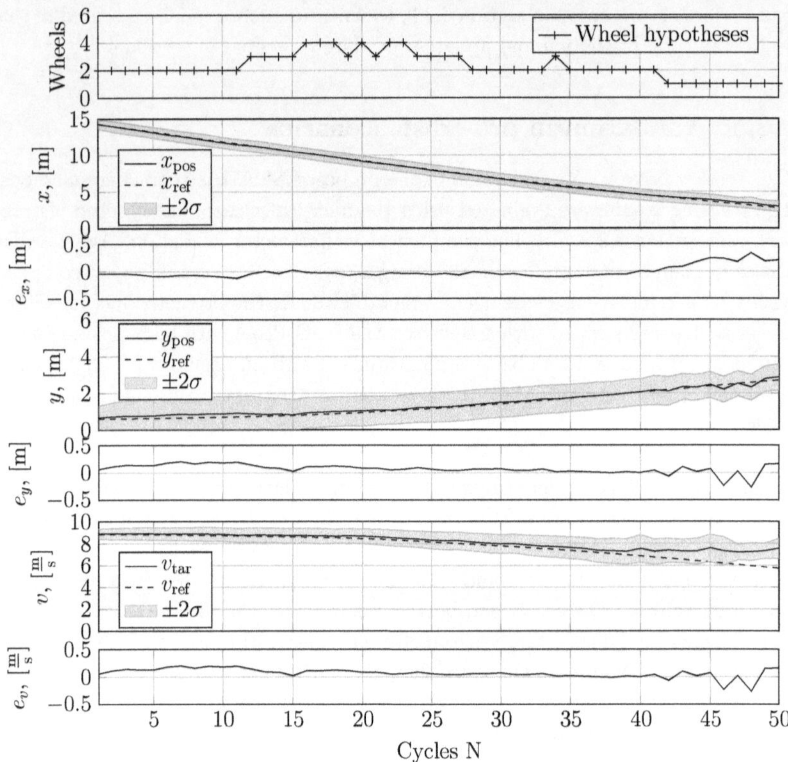

Figure 4.12: SMC-based target state estimation results for x-, y-position and velocity v compared to the true target states measured with a real-time kinematic (RTK) reference system. The top figure shows the number of detected wheel hypotheses per measurement cycle.

The subsequent plots present the values for each x-, y- and v parameter (solid lines), the true target motion (dashed lines) and $\pm 2\sigma$ standard deviation of the estimate for each cycle. Additionally, an error plot for each parameter is presented. The error values for x and y state estimates range between $-0.2\,\mathrm{m}$ and $0.4\,\mathrm{m}$ and the velocity error has its maximum at $0.3\,\mathrm{m/s}$. Notably is the increase of the error values at the end of the measurement where the target is at the FoV edges, therefore by the sensor seen under large angles, and the number of detected wheels decreases.

Fig. 4.13 presents the state estimates for yaw angle ψ, yaw rate w and acceleration a (solid lines) compared to the true target states measured with a RTK reference system. Again, the true target motion (dashed lines) and $\pm 2\sigma$ standard deviation of the estimate for each cycle are indicated and error plots for each parameter are presented. Again, the error for all parameters are comparably low when two or more wheels are detected and increase, notably when the detected wheels drop to one per measurement cycle. The error values for ψ, w and a state estimates range up to $20°$,

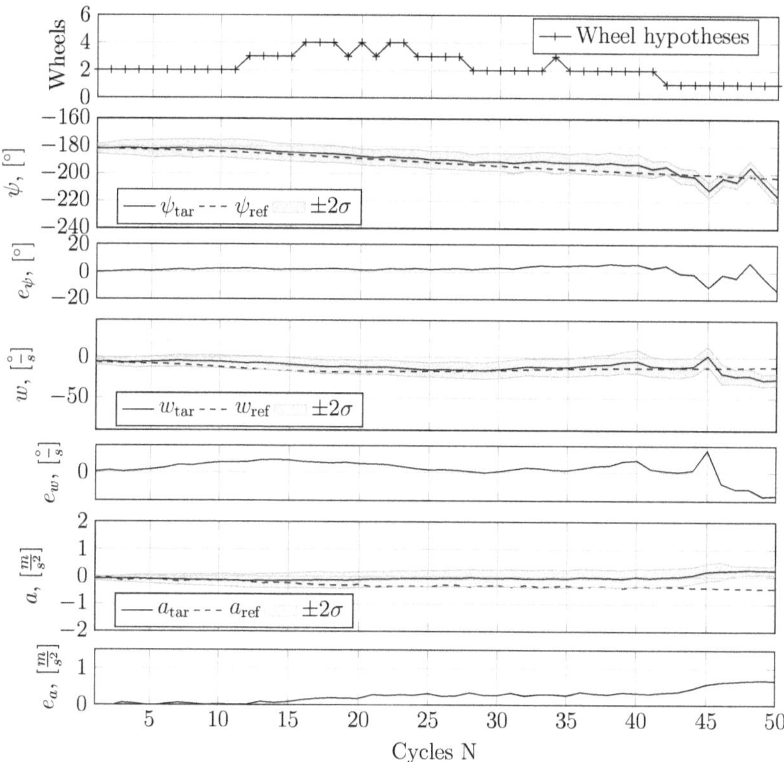

Figure 4.13: SMC-based target state estimation results for yaw angle ψ, yaw rate w and acceleration a compared to the true target states measured with a RTK reference system. The top figure shows the number of detected wheel hypotheses per measurement cycle.

The subsequent plots present the values for each ψ-, w- and a parameter (solid lines), the true target motion (dashed lines) and $\pm 2\sigma$ standard deviation of the estimate for each cycle. Additionally, an error plot for each parameter is presented. The error values for ψ, w and a state estimates range up to $20°$, $40°/s$ and $1.2\,m/s^2$, correspondingly. Notably is the increase of the error values at the end of the measurement where the target is at the FoV edges, therefore by the sensor seen under large angles, and the number of detected wheels decreases.

$40°/s$ and $1.2\,m/s^2$, respectively. At the end of the scenario, only the rear part of the target vehicle is in the sensor FoV and this remaining part is illuminated under large azimuthal angle leading to the decreased tracking performance.

4.3.6 Conclusion

This section presents an extended object tracking framework for vehicles incorporating positions and bulk velocities of detected spinning wheels as a novel procedure. The pre-processing procedure detects the position and corresponding bulk velocities for rotating wheels by evaluating the micro-Doppler effect, which serves as input for the enhanced object tracking procedure. The wheel hypotheses are forwarded to the SMC-based extended object tracking framework to estimate the target dynamic states. The RFS-based measurement model, embedded in the SMC-framework, is capable of incorporating multiple wheel hypotheses backscattered of an extended target object in dynamic driving scenarios, e.g., safety-critical evading maneuvers. The performance was evaluated with real vehicle tests where critical evading driving scenarios are reproducibly reconstructed and results proofed high accuracy for both position and dynamic state estimation.

4.4 Summary

This chapter introduced the principles of target object tracking focused on pre-crash environment perception. The chapter provided an overview of Bayesian state estimation and presented three fundamental filter designs: the Kalman filter (KF), the extended Kalman filter (EKF) for nonlinear state estimation and the particle filter (PF).

Besides theoretical principles, the chapter opened up with a real-world survey evaluating the performance of a series radar sensor. The carrier vehicle is transferred in skid driving scenarios using a novel test method. The vehicle with the mounted radar sensor is transferred in skid scenarios and the radar tracking performance proofed poor accuracy in position and velocity estimation measuring a static radar target.

To enhance the tracking performance in skid driving situations, one of the contributions in this **book** comprises a modified, nonlinear vehicle motion model, which is used to estimate parameters of the horizontal vehicle motion during a skid event.

Subsequently, an extended Kalman filter (EKF) was designed to track the static target object using automotive radar signals as input for this novel filter. The system is tested and evaluated under realistic conditions using a test vehicle equipped with state-of-the-art automotive sensors in skid scenarios. The filter state estimates are compared to the true target motion measured with a reference system and show a significantly improved accuracy than compared to the series sensor as mentioned above.

The next section presented another contribution, a solution to an uncertain measurement model problem, where wandering dominant scatter points on the extended object surface depends on the relative orientation and motion. A subsequent tracking procedure to incorporate target wheel hypotheses from the previous chapter is enfolded. The positions and velocity of spinning target wheels serve as fixed and characteristic points on the target vehicle and are suitable for subsequent tracking. The particle filter (PF)-based tracking framework determines the position and corresponding dynamic parameters for single targets based on wheel hypotheses and is evaluated in real evading maneuvers.

References

[ABBP⁺11] Eva Aigner-Breuss, Christian Brandstaetter, Monika Pilgerstorfer, Anna Müller, and Michael Gatscha. Fahrsicherheitstraining als Massnahme des aktiven Risk Managements. *Forschungsarbeiten des oesterreichischen Verkehrssicherheitsfonds*, 2011.

[AMGC02] Sanjeev M. Arulampalam, Simon Maskell, Neil Gordon, and Tim Clapp. A tutorial on particle filters for online nonlinear/non-Gaussian Bayesian tracking. *IEEE Transactions on signal processing*, 50(2):174–188, 2002.

[BP99] Samuel Blackman and Robert Popoli. Design and analysis of modern tracking systems (Artech House Radar Library). *Artech house*, 1999.

[BSWT11] Yaakov Bar-Shalom, Peter K. Willett, and Xin Tian. *Tracking and data fusion*. YBS publishing Storrs, CT, USA:, 2011.

[BY06] Markus Buhren and Bin Yang. Simulation of automotive radar target lists using a novel approach of object representation. In *2006 IEEE Intelligent Vehicles Symposium*, pages 314–319. IEEE, 2006.

[C⁺03] Zhe Chen et al. Bayesian filtering: From Kalman filters to particle filters, and beyond. *Statistics*, 182(1):1–69, 2003.

[Dau05] Fred Daum. Nonlinear filters: Beyond the Kalman filter. *IEEE Aerospace and Electronic Systems Magazine*, 20(8):57–69, 2005.

[DDFG01] Arnaud Doucet, Nando De Freitas, and Neil Gordon. An introduction to sequential Monte Carlo methods. In *Sequential Monte Carlo methods in practice*, pages 3–14. Springer, 2001.

[DGA00] Arnaud Doucet, Simon Godsill, and Christophe Andrieu. On sequential Monte Carlo sampling methods for Bayesian filtering. *Statistics and computing*, 10(3):197–208, 2000.

[GBR16] Karl Granstrom, Marcus Baum, and Stephan Reuter. Extended object tracking: Introduction, overview and applications. *arXiv preprint arXiv:1604.00970*, 2016.

[GGMS05] Kevin Gilholm, Simon Godsill, Simon Maskell, and David Salmond. Poisson models for extended target and group tracking. In *Signal and Data Processing of Small Targets 2005*, volume 5913, page 59130R. International Society for Optics and Photonics, 2005.

[GLGO14] Karl Granstrom, Christian Lundquist, Fredrik Gustafsson, and Umut Orguner. Random set methods: Estimation of multiple extended objects. *IEEE robotics & automation magazine*, 21(2):73–82, 2014.

[GRA04] Neil Gordon, B. Ristic, and S. Arulampalam. Beyond the Kalman filter: Particle filters for tracking applications. *Artech House, London*, 830:5, 2004.

[GSDB07] Joakim Gunnarsson, Lennart Svensson, Lars Danielsson, and Fredrik Bengtsson. Tracking vehicles using radar detections. In *2007 IEEE Intelligent Vehicles Symposium*, pages 296–302. IEEE, 2007.

[GSS93] Neil J. Gordon, David J. Salmond, and Adrian F.M. Smith. Novel approach to nonlinear/non-Gaussian Bayesian state estimation. In *IEE proceedings F (radar and signal processing)*, volume 140, pages 107–113. IET, 1993.

[HLS12] Lars Hammarstrand, Malin Lundgren, and Lennart Svensson. Adaptive radar sensor model for tracking structured extended objects. *IEEE Transactions on Aerospace and Electronic Systems*, 48(3):1975–1995, 2012.

[HSSS12] Lars Hammarstrand, Lennart Svensson, Fredrik Sandblom, and Joakim Sorstedt. Extended object tracking using a radar resolution model. *IEEE Transactions on Aerospace and Electronic Systems*, 48(3):2371–2386, 2012.

[Kal60] Rudolph E. Kalman. A new approach to linear filtering and prediction problems. *Journal of basic Engineering*, 82(1):35–45, 1960.

[Kam63] Wunibald Kamm. *Modellversuche und Berechnungen über die Richtungshaltung von Kraftfahrzeugen*. VDI-Verlag, 1963.

[KBH+17] Alexander Kamann, Jonas B. Bielmeier, Sinan Hasirlioglu, Ulrich T. Schwarz, and Thomas Brandmeier. Object tracking based on an extended Kalman filter in high dynamic driving situations. In *2017 IEEE 20th International Conference on Intelligent Transportation Systems (ITSC)*, pages 1–6. IEEE, 2017.

[KBK+16] Dominik Kellner, Michael Barjenbruch, Jens Klappstein, Jürgen Dickmann, and Klaus Dietmayer. Tracking of extended objects with high-resolution Doppler radar. *IEEE Transactions on Intelligent Transportation Systems*, 17(5):1341–1353, 2016.

[KHD+17] Alexander Kamann, Sinan Hasirlioglu, Igor Doric, Thomas Speth, Thomas Brandmeier, and Ulrich T. Schwarz. Test methodology for automotive surround sensors in dynamic driving situations. In *2017 IEEE 85th Vehicular Technology Conference (VTC Spring)*, pages 1–6. IEEE, 2017.

[KSD16] Christina Knill, Alexander Scheel, and Klaus Dietmayer. A direct scattering model for tracking vehicles with high-resolution radars. In *2016 IEEE Intelligent Vehicles Symposium (IV)*, pages 298–303. IEEE, 2016.

[LJ03] Rong X. Li and Vesselin P. Jilkov. Survey of maneuvering target tracking. part I. dynamic models. *IEEE Transactions on aerospace and electronic systems*, 39(4):1333–1364, 2003.

[Mah07] Ronald P.S. Mahler. *Statistical multisource-multitarget information fusion*. Artech House, Inc., 2007.

[Pac66] Hans Bastiaan Pacejka. *The wheel shimmy phenomenon: a theoretical and experimental investigation with particular reference to the non-linear problem*. PhD book, TU Delft, Delft University of Technology, 1966.

[Pac05] Hans Pacejka. *Tire and vehicle dynamics*. Elsevier, 2005.

[RS40] Paul Riekert and Theo-Ernst Schunck. Zur Fahrmechanik des gummibereiften Kraftfahrzeugs. *Ingenieur-Archiv*, 11(3):210–224, 1940.

[RS04] Branko Ristic and David J. Salmond. A study of a nonlinear filtering problem for tracking an extended target. In *Proccedings on 7th International Conference on Information Fusion*, pages 503–509. Citeseer, 2004.

[San13] Volker Sandner. Development of a test target for aeb systems. In *23rd International Technical Conference on the Enhanced Safety of Vehicles (ESV): Research Collaboration to Benefit Safety of All Road Users*, 2013.

[Sch14] Dieter Schramm. *Vehicle dynamics : modeling and simulation*. Springer, 2014.

[SD18] Alexander Scheel and Klaus Dietmayer. Tracking multiple vehicles using a variational radar model. In *IEEE Transactions on Intelligent Transportation Systems*. IEEE, 2018.

[SKRD16] Alexander Scheel, Christina Knill, Stephan Reuter, and Klaus Dietmayer. Multi-sensor multi-object tracking of vehicles using high-resolution radars. In *2016 IEEE Intelligent Vehicles Symposium (IV)*, pages 558–565. IEEE, 2016.

[SP03] David J. Salmond and M.C. Parr. Track maintenance using measurements of target extent. *IEE Proceedings-Radar, Sonar and Navigation*, 150(6):389–395, 2003.

[SRW08] Robin Schubert, Eric Richter, and Gerd Wanielik. Comparison and evaluation of advanced motion models for vehicle tracking. In *2008 11th international conference on information fusion*, pages 1–6. IEEE, 2008.

[TBF05] Sebastian Thrun, Wolfram Burgard, and Dieter Fox. *Probabilistic robotics*. MIT press, 2005.

[YGMB07] Chun Yang, Wendy Garber, Richard Mitchell, and Erik Blasch. A simple maneuver indicator from target's range-Doppler image. In *2007 10th International Conference on Information Fusion*, pages 1–8. IEEE, 2007.

Chapter 5

Conclusions and outlook

5.1   Conclusions

In this book, novel techniques for radar-based target detection and tracking in dynamic pre-crash scenarios, as well as for ghost object identification, have been developed. This chapter summarizes the scientific contributions, emphasizes their practical value for the proposed approaches and provides an outlook for subsequent future work.

First, the two-fold problem of an uncertain measurement model due to a wandering dominant scatter point on the target surface and corresponding challenge for accurate target tracking in low-range configurations has been considered. The proposed method presents a procedure to estimate target wheel positions and corresponding bulk velocities based on radar sensor data. The scientific contribution is to spatially resolve the micro-Doppler signals, generated by the rotating wheels of the target vehicle, to determine characteristic scatter points of the target. As a consequence, the effect of wandering dominant scatter point on the surface of extended objects is mitigated. A micro-Doppler parameter is defined, which quantifies detections that are with high probability generated by the rotating target wheels. These detections are processed to estimate the wheel position and corresponding bulk velocities of the target, referred to as wheel hypotheses. The proposed method is evaluated in dynamic driving scenarios, where the driver performs an emergency evading action to avoid a collision. The results proof feasibility of the proposed method and enable extended object tracking, incorporating the novel target information. Subsequently, a sequential Monte Carlo (SMC)-based extended object tracking frame-work for vehicles incorporating positions and bulk velocities of detected spinning wheels procedure is developed. The wheel hypotheses are forwarded to the sequen-tial Monte Carlo (SMC)-based extended object tracking framework to estimate the target position and dynamic states. The random finite set (RFS)-based measure-ment model, embedded in the sequential Monte Carlo (SMC)-framework, is capable of incorporating multiple wheel hypotheses backscattered of an extended target ob-ject in a high dynamic driving scenario, e.g., safety-critical evading maneuvers. The tracking performance was evaluated in real vehicle test runs, where critical evading driving scenarios are reproducibly reconstructed and results proofed high accuracy for both position and dynamic state estimation.

Second, the **book** emphasized the ghost object generation problem due to mul-tipath propagation in pre-crash scenarios. Radar sensors, perceiving the immediate lateral vehicle environment, show an elevated ghost object presence due to a higher probability illuminating potential reflection surfaces, e.g., road boundaries or build-ings. At times, these ghost objects appear to be on a collision trajectory with the ego vehicle, whereas the vehicles are in uncritical driving scenarios, e.g., an urban inter-section. One target object may generate multiple false-positive targets in real-world driving scenarios. Based on the propagation and reflection behavior of electromag-netic waves, a geometric multipath model is derived, illustrating the occurring mul-tipath reflections on real-world surfaces, e.g., buildings or road-bounding barriers. The proposed geometric propagation model describes the relative positions of the false-positive reflections and is validated with extensive real radar sensor data. A custom reflector target mounted on a platform, creating deterministic point targets as dominant backscatter centers of a vehicle body, validated the different multipath reflections and the overall accuracy of the model. Moreover, radar measurements of a vehicle during an intersection scenario prooved relevance to multipath reflection behavior and confirmed the model assumptions. The results identify the emerging challenges for the correct interpretation of false-positive targets to enable sophisti-cated automotive safety-, comfort- and automation applications.

Third, the relevance of skid scenarios with high magnitudes of side slip angles in pre-crash phases is highlighted and a tracking procedure is developed. Therefore, a novel test methodology, to non-destructively transfer vehicles with mounted surround sen-sors in skid situations, is developed and a survey analyzing a state-of-the-art radar sensor revealed object tracking improvement potential. A test vehicle, equipped with a state-of-the-art automotive radar sensor and a reference sensor, was tested in real skid situations using a kick plate and a standardized radar target. The method utilizes the side slip angle as a criticality criterion, which may be adjusted by the kick plate. Subsequently, a modified motion model is derived, estimating side slip angles in these skid driving situations and serves as input for an extended Kalman filter (EKF)-based target tracking procedure. The contribution emphasizes the es-timation of horizontal vehicle motion using the proposed modified, nonlinear single track model considering an additional lateral force applied to the vehicle rear axle. Based on these results, an extended Kalman filter is designed estimating the target object relative position and velocity in skid scenarios and validated both, the track-ing and side slip angle estimations in real car tests using the above-mentioned test method.

5.2 Outlook

This section discusses briefly possible future work.

Spatially resolved micro-Doppler signals and tracking: The procedure to determine target wheel positions and corresponding bulk velocities perform well with real data from proposed measurements. As a next stage, real-time implementation and extensive validation with experimental data, covering a wide range of pre-crash sensor-to-target configurations, seem appropriate. Procedure improvement possibilities may be realized by the application of sophisticated estimation methods for data segmentation and processing, e.g., machine learning [LHDW16, PRV+19, SHDW18]. Additionally, a wheel labeling procedure to label each wheel seems beneficial for future tracking states.

Target tracking in skid scenarios: Based on the achieved performance, the next EKF development stage is an extension for dynamic target object tracking. The modified motion model is suitable to map high yaw rates and side slip angles real-world relevant in, e.g., multi-collision scenarios. Alternative model-based side slip angle estimations [CH08, YHL09] may serve as additional input for the tracking procedure to further enhance tracking accuracy.

Multipath propagation in uncertain environments: Further investigations on multipath reflections for automotive radar sensors are carried out to analyze the influence also on the object velocity estimation as well as extend the target types also by vulnerable road users, e.g., pedestrians. Observations and literature proof a shifted velocity signal for each false-positive object [BMS+18, VHZ18].

References

[BMS+18] Vladimir Brazda, Michal Mandlik, Christian Sturm, Stefan Görner, and Urs Lübbcrt. Automotive radar sensor-wave propagation issues and their mitigation. In *2018 International Symposium ELMAR*, pages 135–138. IEEE, 2018.

[CH08] Bo C. Chen and Chang F. Hsieh. Sideslip angle estimation using extended Kalman filter. *Vehicle System Dynamics*, 46(S1):353–364, 2008.

[LHDW16] Jakob Lombacher, Markus Hahn, Jürgen Dickmann, and Christian Wöhler. Potential of radar for static object classification using deep learning methods. In *2016 IEEE MTT-S International Conference on Microwaves for Intelligent Mobility (ICMIM)*, pages 1–4. IEEE, 2016.

[PRV+19] Kanil Patel, Kilian Rambach, Tristan Visentin, Daniel Rusev, Michael Pfeiffer, and Bin Yang. Deep learning-based object classification on automotive radar spectra. In *IEEE Radar Conference, Boston, MA, USA*, pages 22–26, 2019.

[SHDW18] Ole Schumann, Markus Hahn, Jürgen Dickmann, and Christian Wöhler. Semantic segmentation on radar point clouds. In *2018 21st International Conference on Information Fusion (FUSION)*, pages 2179–2186. IEEE, 2018.

[VHZ18] Tristan Visentin, Jürgen Hasch, and Thomas Zwick. Analysis of multipath and DOA detection using a fully polarimetric automotive radar. *International Journal of Microwave and Wireless Technologies*, 10(5-6):570–577, 2018.

[YHL09] Seung-Han You, Jin-Oh Hahn, and Hyeongcheol Lee. New adaptive approaches to real-time estimation of vehicle sideslip angle. *Control Engineering Practice*, 17(12):1367–1379, 2009.